T0301821

Materials and Physics for Nonvolatile Memories II

MATERIALS RESEARCH SOCIETY
SYMPOSIUM PROCEEDINGS VOLUME 1250

Materials and Physics for Nonvolatile Memories II

Spring 2010, April 5–9, San Francisco, California, U.S.A.

EDITORS:

Caroline Bonafos
CEMES/CNRS
Toulouse, France

Yoshihisa Fujisaki
Hitachi Ltd.
Kokubunji, Tokyo, Japan

Panagiotis Dimitrakis
IMEL/NCSR "Demokritos"
Aghia Paraskevi, Greece

Eisuke Tokumitsu
Tokyo Institute of Technology
Yokohama, Japan

Materials Research Society
Warrendale, Pennsylvania

CAMBRIDGE
UNIVERSITY PRESS

Shaftesbury Road, Cambridge CB2 8EA, United Kingdom

One Liberty Plaza, 20th Floor, New York, NY 10006, USA

477 Williamstown Road, Port Melbourne, VIC 3207, Australia

314–321, 3rd Floor, Plot 3, Splendor Forum, Jasola District Centre, New Delhi – 110025, India

103 Penang Road, #05–06/07, Visioncrest Commercial, Singapore 238467

Cambridge University Press is part of Cambridge University Press & Assessment, a department of the University of Cambridge.

We share the University's mission to contribute to society through the pursuit of education, learning and research at the highest international levels of excellence.

www.cambridge.org
Information on this title: www.cambridge.org/9781605112275

Materials Research Society
506 Keystone Drive, Warrendale, PA 15086
http://www.mrs.org

First published 2010
First paperback edition 2012

Single article reprints from this publication are available through University Microfilms Inc., 300 North Zeeb Road, Ann Arbor, MI 48106

A catalogue record for this publication is available from the British Library

CODEN: MRSPDH

ISBN   978-1-605-11227-5   Hardback
ISBN   978-1-107-40799-2   Paperback

CONTENTS

ADVANCED FLASH MEMORY

NANO-CRYSTAL MEMORY

*Invited Paper

MAGNETORESISTIVE RAM (MRAM)

FERROELECTRIC RAM (FeRAM)

*Invited Paper

*Invited Paper

RESISTANCE RAM (RERAM)

PREFACE

Symposium G, "Materials and Physics of Nonvolatile Memories" was held April 5–9 at the 2010 MRS Spring Meeting in San Francisco, California. This is the fourth in a series of MRS symposia on nonvolatile memories. About 120 papers were presented, including 12 invited presentations, indicating the importance of this technological field. The symposium, which covered scientific and technological exploration of materials and physics of various nonvolatile memory devices, served as a forum for scientists and engineers on different nonvolatile memory technologies to discuss the latest technical advances and future trends. Nonvolatile memory technologies and related materials issues discussed at the symposium were advanced Flash memories, semiconducting and metallic nanocrystal memories, magnetoresistive RAM (MRAM), ferroelectric RAM (FeRAM), organic memories, phase-change RAM, and resistive switching RAM (ReRAM).

This symposium proceedings volume represents the latest technical advances and related materials issues on many kinds of nonvolatile memory technologies. The papers in this volume are organized according to each type of memory technology and are not in the order of the symposium presentations. The volume also includes some papers from a joint session between this symposium and Symposium H, "Phase-Change Materials for Memory and Reconfigurable Electronics Applications."

We would like to thank all of the speakers and participants for their valuable contributions toward making the symposium successful. We gratefully acknowledge the financial support of Annealsys, CNRS, Park Systems Corp., Universal Systems Corp., WACOM R&D Corp.

Caroline Bonafos
Yoshihisa Fujisaki
Panagiotis Dimitrakis
Eisuke Tokumitsu

August 2010

MATERIALS RESEARCH SOCIETY SYMPOSIUM PROCEEDINGS

Volume 1245 — Amorphous and Polycrystalline Thin-Film Silicon Science and Technology—2010, Q. Wang, B. Yan, C.C. Tsai, S. Higashi, A. Flewitt, 2010, ISBN 978-1-60511-222-0

Volume 1246 — Silicon Carbide 2010—Materials, Processing and Devices, S.E. Saddow, E.K. Sanchez, F. Zhao, M. Dudley, 2010, ISBN 978-1-60511-223-7

Volume 1247E —Solution Processing of Inorganic and Hybrid Materials for Electronics and Photonics, 2010, ISBN 978-1-60511-224-4

Volume 1248E —Plasmonic Materials and Metamaterials, J.A. Dionne, L.A. Sweatlock, G. Shvets, L.P. Lee, 2010, ISBN 978-1-60511-225-1

Volume 1249 — Advanced Interconnects and Chemical Mechanical Planarization for Micro- and Nanoelectronics, J.W. Bartha, C.L. Borst, D. DeNardis, H. Kim, A. Naeemi, A. Nelson, S.S. Papa Rao, H.W. Ro, D. Toma, 2010, ISBN 978-1-60511-226-8

Volume 1250 — Materials and Physics for Nonvolatile Memories II, C. Bonafos, Y. Fujisaki, P. Dimitrakis, E. Tokumitsu, 2010, ISBN 978-1-60511-227-5

Volume 1251E —Phase-Change Materials for Memory and Reconfigurable Electronics Applications, P. Fons, K. Campbell, B. Cheong, S. Raoux, M. Wuttig, 2010, ISBN 978-1-60511-228-2

Volume 1252— Materials and Devices for End-of-Roadmap and Beyond CMOS Scaling, A.C. Kummel, P. Majhi, I. Thayne, H. Watanabe, S. Ramanathan, S. Guha, J. Mannhart, 2010, ISBN 978-1-60511-229-9

Volume 1253 — Functional Materials and Nanostructures for Chemical and Biochemical Sensing, E. Comini, P. Gouma, G. Malliaras, L. Torsi, 2010, ISBN 978-1-60511-230-5

Volume 1254E —Recent Advances and New Discoveries in High-Temperature Superconductivity, S.H. Wee, V. Selvamanickam, Q. Jia, H. Hosono, H-H. Wen, 2010, ISBN 978-1-60511-231-2

Volume 1255E —Structure-Function Relations at Perovskite Surfaces and Interfaces, A.P. Baddorf, U. Diebold, D. Hesse, A. Rappe, N. Shibata, 2010, ISBN 978-1-60511-232-9

Volume 1256E —Functional Oxide Nanostructures and Heterostructures, 2010, ISBN 978-1-60511-233-6

Volume 1257 — Multifunctional Nanoparticle Systems—Coupled Behavior and Applications, Y. Bao, A.M. Dattelbaum, J.B. Tracy, Y. Yin, 2010, ISBN 978-1-60511-234-3

Volume 1258 — Low-Dimensional Functional Nanostructures—Fabrication, Characterization and Applications, H. Riel, W. Lee, M. Zacharias, M. McAlpine, T. Mayer , H. Fan, M. Knez, S. Wong, 2010, ISBN 978-1-60511-235-0

Volume 1259E —Graphene Materials and Devices, M. Chhowalla, 2010, ISBN 978-1-60511-236-7

Volume 1260 — Photovoltaics and Optoelectronics from Nanoparticles, M. Winterer, W.L. Gladfelter, D.R. Gamelin, S. Oda, 2010, ISBN 978-1-60511-237-4

Volume 1261E —Scanning Probe Microscopy—Frontiers in NanoBio Science, C. Durkan, 2010, ISBN 978-1-60511-238-1

Volume 1262 — In-Situ and Operando Probing of Energy Materials at Multiscale Down to Single Atomic Column—The Power of X-Rays, Neutrons and Electron Microscopy, C.M. Wang, N. de Jonge, R.E. Dunin-Borkowski, A. Braun, J-H. Guo, H. Schober, R.E. Winans, 2010, ISBN 978-1-60511-239-8

Volume 1263E —Computational Approaches to Materials for Energy, K. Kim, M. van Shilfgaarde, V. Ozolins, G. Ceder, V. Tomar, 2010, ISBN 978-1-60511-240-4

Volume 1264 — Basic Actinide Science and Materials for Nuclear Applications, J.K. Gibson, S.K. McCall, E.D. Bauer, L. Soderholm, T. Fanghaenel, R. Devanathan, A. Misra, C. Trautmann, B.D. Wirth, 2010, ISBN 978-1-60511-241-1

Volume 1265 — Scientific Basis for Nuclear Waste Management XXXIV, K.L. Smith, S. Kroeker, B. Uberuaga, K.R. Whittle, 2010, ISBN 978-1-60511-242-8

Volume 1266E —Solid-State Batteries, S-H. Lee, A. Hayashi, N. Dudney, K. Takada, 2010, ISBN 978-1-60511-243-5

Volume 1267 — Thermoelectric Materials 2010—Growth, Properties, Novel Characterization Methods and Applications, H.L. Tuller, J.D. Baniecki, G.J. Snyder, J.A. Malen, 2010, ISBN 978-1-60511-244-2

Volume 1268 — Defects in Inorganic Photovoltaic Materials, D. Friedman, M. Stavola, W. Walukiewicz, S. Zhang, 2010, ISBN 978-1-60511-245-9

Volume 1269E —Polymer Materials and Membranes for Energy Devices, A.M. Herring, J.B. Kerr, S.J. Hamrock, T.A. Zawodzinski, 2010, ISBN 978-1-60511-246-6

MATERIALS RESEARCH SOCIETY SYMPOSIUM PROCEEDINGS

Volume 1270 — Organic Photovoltaics and Related Electronics—From Excitons to Devices,
V.R. Bommisetty, N.S. Sariciftci, K. Narayan, G. Rumbles, P. Peumans, J. van de Lagemaat,
G. Dennler, S.E. Shaheen, 2010, ISBN 978-1-60511-247-3
Volume 1271E —Stretchable Electronics and Conformal Biointerfaces, S.P. Lacour, S. Bauer, J. Rogers,
B. Morrison, 2010, ISBN 978-1-60511-248-0
Volume 1272 — Integrated Miniaturized Materials—From Self-Assembly to Device Integration,
C.J. Martinez, J. Cabral, A. Fernandez-Nieves, S. Grego, A. Goyal, Q. Lin, J.J. Urban,
J.J. Watkins, A. Saiani, R. Callens, J.H. Collier, A. Donald, W. Murphy, D.H. Gracias,
B.A. Grzybowski, P.W.K. Rothemund, O.G. Schmidt, R.R. Naik, P.B. Messersmith,
M.M. Stevens, R.V. Ulijn, 2010, ISBN 978-1-60511-249-7
Volume 1273E —Evaporative Self Assembly of Polymers, Nanoparticles and DNA , B.A. Korgel, 2010,
ISBN 978-1-60511-250-3
Volume 1274 — Biological Materials and Structures in Physiologically Extreme Conditions and Disease,
M.J. Buehler, D. Kaplan, C.T. Lim, J. Spatz, 2010, ISBN 978-1-60511-251-0

Prior Materials Research Society Symposium Proceedings available by contacting Materials Research Society

Advanced Flash Memory

Mater. Res. Soc. Symp. Proc. Vol. 1250 © 2010 Materials Research Society   1250-G07-03

Charge trapping memories with atomic layer deposited high-k dielectrics capping layers

N. Nikolaou[1], P. Dimitrakis[1], P. Normand[1], K. Giannakopoulos[2], K. Mergia[3], V. Ioannou-Sougleridis[1], Kaupo Kukli [4,5], Jaakko Niinistö[4], Mikko Ritala[4] and Markku Leskelä[4]
[1] Institute of Microelectronics, NSCD "Demokritos" 153-10, Athens, Greece,
[2] Institute of Materials Science, NSCD "Demokritos" 153-10 Athens, Greece,
[3] Institute of Nuclear Technology and Rad. Prot. NSCD "Demokritos" 153-10 Athens, Greece,
[4] Department of Chemistry, University of Helsinki FI 00014, Finland,
[5] Institute of Physics, University of Tartu, Estonia

ABSTRACT

In this work, we examine the influence of hafnium and zirconium oxides ALD precursor chemistry on the memory properties of $SiO_2/Si_3N_4/ZrO_2$ and $SiO_2/Si_3N_4/HfO_2$ non-volatile gate memory stacks. Approximately 10 nm thick ZrO_2 and HfO_2 layers were deposited on top of a SiO_2/Si_3N_4 structure, functioning as blocking oxides. Both metal oxides were deposited using either alkylamides or cyclopentadienyls as metal precursors, and ozone as the oxygen source. In the case of the ZrO_2 gate stacks a memory window of 6 V was determined, comprised of 4 V write window and 2 V erase window. Although no dramatic/major differences were evident between the ZrO_2 layers, ZrO_2 grown from alkylamide provided structures with higher dielectric strength. The memory structures with HfO_2 blocking layers indicate that the memory window and the dielectric strength are significantly affected by the precursor. The structures with the HfO_2 formed from alkylamide showed a write window of 7 V, while the films grown from cyclopentadienyl possessed window of 5 V. Comparison between the memory windows obtained using ZrO_2 and HfO_2 as control oxides reveals that the former provides memory structures with higher electron trapping efficiency.

INTRODUCTION

The continuation of scaling of the standard floating gate non-volatile memory devices faces many difficult challenges, such as further reduction of the control and the tunnel oxides thicknesses which result in charge leakage [1]. In addition, the close proximity of the adjacent devices leads also to floating gate interference effects. Silicon nitride charge trapping technology offers an alternative scaling route, provided that the over-erase effect will be circumvented [2]. This approach implies the introduction of high permittivity dielectric layers that will replace either one or more of the constituent layers of the oxide-nitride-oxide typical stack gate [3]. The replacement of the dielectric layers is usually combined with high work function metal gates in order to suppress electron injection from the gate, during the erase operation [4]. Atomic Layer Deposition (ALD) can be regarded as a deposition method preferred for the growth of high-k dielectric layers. A critical aspect of this technology is the choice of precursors, i.e. the factor determining the quality and the properties of the deposited layer [5,6].

EXPERIMENT

Silicon substrates (n-type, 1-2 Ω cm) were first oxidized to a thickness of 2.5 nm at 800 °C in N_2O ambient, followed by low pressure chemical vapor deposition of a 5 nm Si_3N_4 layer at 800 °C (NH_3, dichlorosilane). On top of the oxide-nitride stack 10 nm thick layers of hafnium or zirconium oxides were grown by ALD. Hafnium oxide was formed using either the hafnium tetrakis(ethylmethylamide), $Hf[N(C_2H_5)CH_3)]_4$, and ozone (O_3) at 275 °C, or using the cyclopentadienyl precursor $((CpMe)_2Hf(OMe)Me, Cp = C_5H_5, Me = CH_3)$ and ozone at 350 °C. Analogous processes were used for the deposition of the zirconium oxide layers, i.e. $Zr[N(C_2H_5)CH_3)]_4$ and O_3 at 275 °C, and $(CpMe)_2Zr(OMe)Me$ and O_3 at 350 °C. The processes were selected due to the good quality of the deposited layers and low amounts of residual impurities [5,6]. The effect of the oxygen annealing on the gate stack was also studied and parts of the samples were subjected to annealing at 600 °C for 2 min in oxygen ambient. Platinum electrode MOS capacitors were fabricated using photolithography and lift-off process in order to study the basic electrical characteristics (current-voltage and capacitance-voltage) as well as the memory properties of the gate stack. The area of the capacitors under investigation was 1×10^{-4} cm^{-2}.

RESULTS
Structural characteristics

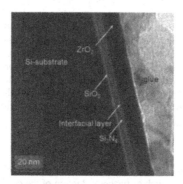

Figure1. Dark field cross-sectional TEM image of a $SiO_2/Si_3N_4/ZrO_2$ dielectric stack with the high-k ZrO_2 layer deposited in the cyclopentadienyl/O_3 process.

Fig.1 shows a typical transmission electron microscopy (TEM) cross-section image of a $SiO_2/Si_3N_4/ZrO_2$ structure where the ZrO_2 was deposited by the cyclopentadienyl/O_3 process on the free surface of the Si_3N_4 layer. The SiO_2 layer thickness was estimated to be around 2.5 nm, the Si_3N_4 layer thickness 5 nm and the ZrO_2 layer 10 nm. The interface between the ZrO_2 and Si_3N_4 is rough and the contrast indicates the formation of a 0.5 nm thick interfacial layer, probably due to the influence of ozone on the Si_3N_4. In addition, the plan view TEM image reveals that the ZrO_2 layer deposited using cyclopentadienyl and O_3 precursors is polycrystalline (Fig.2). In addition, the plan view TEM image reveals that the deposited ZrO_2 layer using the cyclopentadienyl/O_3 precursors is polycrystalline (Fig.2). Grazing incidence X-Ray diffraction (GIXRD) studies (not shown) reveal that ZrO_2 layers deposited with both ALD chemistries crystallize in the tetragonal system (Space Group: P42/nmc) with an average crystallite size of 14 nm as determined from profile analysis of the GIXRD spectra considering the instrumental resolution. Annealing increases by about 6% the size of the crystallites. The GIXRD spectra of

4

the HfO$_2$ films deposited by the alkylamide/O$_3$ process present two very broad Bragg peaks at 32 and 55.5 degrees corresponding to a mean grain size of about 2 nm determined using the Scherrer equation. For the case of the cyclopentadienyl/O$_3$ process the HfO$_2$ layer crystallizes in the monoclinic system (SG: P21/a) while the average crystallite size is about 10 nm.

Figure 2. Plan view TEM bright-field image of a SiO$_2$/Si$_3$N$_4$/ZrO$_2$ dielectric stack with the high-k ZrO$_2$ layer deposited by the cyclopentadienyl/O$_3$ process. The image shows numerous ZrO$_2$ crystallites which compose the layer.

<u>C-V and I-V characteristics</u>

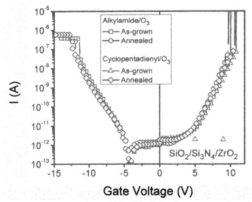

Figure 3. I-V characteristics of the SiO$_2$/Si$_3$N$_4$/ZrO$_2$, gate stacks, deposited from cyclopentadienyl and alkylamide precursors, before and after annealing.

Fig. 3 shows typical I-V characteristics of the SiO$_2$/Si$_3$N$_4$/ZrO$_2$ gate stacks with Pt gate electrodes. The characteristics were recorded by scanning the gate voltage from 0 V towards positive or negative gate voltages with a step of 0.1 V/s. The measurements at inversion (negative bias regime) were obtained under white light illumination to provide minority carriers at the substrate. Conduction at accumulation takes place via injection of substrate electrons and starts at 5 V, which corresponds to an average electric field of 6.5 MV/cm across the dielectric stack. Conduction at the inversion regime is initiated by the injection of substrate holes, while at voltages higher than -10 V injection of electrons from the gate electrode occurs simultaneously. The I-V characteristics indicate no significant differences between the examined gate stacks outside the statistical variation range. Thus, the two different ALD chemistries which were employed for the deposition of the ZrO_2 control oxides provide structures with similar conduction characteristics which are not influenced by annealing. In addition, characteristics similar for films grown from both precursors were obtained from the capacitance-voltage measurements (not shown). The gate stack capacitance at accumulation for the films grown from cyclopentadienyl and O_3 is approximately 4.7×10^{-7} F/cm^2, while that for the films grown from alkylamide and O_3 is 4.8×10^{-7} F/cm^2, both measured at a frequency of 1 MHz. It is well known that the accumulation capacitance of the gate stack is the series capacitance combination of each dielectric layer according to equation (1).

$$\frac{1}{C_{acc}} = \frac{t_{ox}}{\varepsilon_{ox}} + \frac{t_{nit}}{\varepsilon_{nit}} + \frac{t_{ZrO_2}}{\varepsilon_{ZrO_2}} + \frac{t_{int}}{\varepsilon_{int}} \quad (1)$$

Here, t_{ox} and ε_{ox} are the thickness and the dielectric constant of SiO_2 layer, respectively, t_{nit} and ε_{nit} those of the Si_3N_4, t_{ZrO2}, ε_{ZrO2} those of the zirconium oxide layer, and t_{int} and ε_{int} are those of the interfacial layer. Using the well known values of oxide and nitride dielectric constants, the known values of the layer thicknesses, and assuming an intermediate value of 5 for the dielectric constant of the interfacial layer it is possible to estimate the dielectric constant of the deposited ZrO_2, as well as the capacitance equivalent oxide thickness (CET) of the gate stack. For the 11 nm thick film grown from cyclopentadienyl and O_3, ε_{ZrO2} was found to be around 19 with CET around 7.5 nm. For the 10 nm thick film grown from alkylamide and O_3, the dielectric constant was estimated around 17.5 with CET of 7.2 nm.

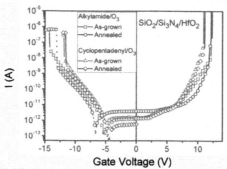

Figure 4. I-V characteristics of the $SiO_2/Si_3N_4/HfO_2$, gate stacks, deposited from cyclopentadienyl/O_3 and alkylamide/O_3 precursors, before and after annealing.

The corresponding I-V characteristics of the $SiO_2/Si_3N_4/HfO_2$ gate stacks with Pt gate electrode are presented in Fig. 4. The differences of the I-V characteristics between these gate stacks are probably caused by the differences in ALD process chemistry. Electron injection set-in for the films grown from alkylamide and O_3 occurs at an average electric field of 7.6 MV/cm, while for the films grown from cyclopentadienyl and O_3 it slightly increases to 8 MV/cm. The effective dielectric constant ε_{HfO2} was estimated at 14.6 for the films grown using alkylamide chemistry and 15.3 for the cyclopentadienyl. After annealing these values were found to be slightly reduced.

The trapping characteristics of the $SiO_2/Si_3N_4/ZrO_2$ gate stacks are given in Fig. 5 which shows the flat-band voltage shift (ΔV_{FB}) as a function of the applied gate voltage pulse. In these experiments, the pulse duration was fixed at 100 ms and C-V measurements were performed after the application of a single positive (negative) pulse without modifying (charging or discharging) the memory state between the pulses. Electron injection and trapping occurs above 4 V and the resulting ΔV_{FB} increases with the applied voltage. The electron trapping results in a write window of 4 V. Note that the gate stacks with ZrO_2 formed using the alkylamide/O_3 chemistry can withstand larger gate voltages than those grown by the cyclopentadienyl/O_3 process. Hole injection and trapping initiate above 5 V and result in an erase window of -2 V. Note also that the electron injection from the gate during the application of high negative amplitude pulses limits the hole trapping effect and a turn-around tendency in the ΔV_{FB} is clearly manifested. This over-erase effect sets in at lower voltages for the case of film grown from cyclopentadienyl and O_3. At gate pulses above 10-12 V severe degradation of the structure occurs, mainly due to degradation of the control oxide.

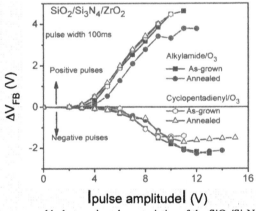

Figure 5. Electron and hole trapping characteristics of the $SiO_2/Si_3N_4/ZrO_2$ structures. Flat band voltage shifts as a function of the gate pulse voltage for 100 ms duration.

The trapping characteristics of the $SiO_2/Si_3N_4/HfO_2$ gate stacks, are presented in Fig. 6. The write window due to electron trapping is around 7 V for the films grown by the alkylamide/O_3 process and 5 V for the films grown from cyclopentadienyl/O_3. Respectively the erase windows due to hole trapping are -3 V and -2 V. The better charge trapping ability of the gate stacks obtained from the alkylamide is also manifested during the application of high amplitude negative pulses, where the cyclopentadienyl-processed gate stacks suffer from the higher gate electron injection, limiting thus the hole trapping window.

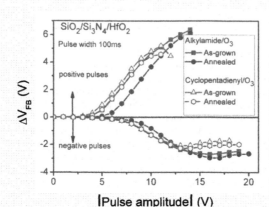

|Pulse amplitude| (V)

Figure 6. Trapping characteristics of the $SiO_2/Si_3N_4/HfO_2$ shown as a flat-band voltage shifts with the gate pulse amplitude. Pulse duration 100 ms.

CONCLUSIONS

The influence of two different precursors on the structural and electrical properties of ZrO_2 and HfO_2 deposited on top of a SiO_2/Si_3N_4 stacks was investigated. In the case of $SiO_2/Si_3N_4/ZrO_2$ stacks similar electrical and structural characteristics were revealed. On the contrary, analytical structural characterization suggested significant differences between the hafnium oxide precursors. However, the electrical characterization showed similar behavior in terms of conduction properties. The memory characteristics of the $SiO_2/Si_3N_4/ZrO_2$ stacks prepared by the alkylamide precursor exhibited slightly larger memory window for positive pulses and higher immunity to the over-erase effect for the negative pulses. Similar effects were observed for the alkylamide deposited HfO_2. Comparison between the memory windows obtained using ZrO_2 and HfO_2 as control oxides reveals that the former provides memory structures with higher electron trapping efficiency.

REFERENCES

[1] C.-Y. Lu, K.-Y. Hsieh, and R. Liu, Microelectronics Engineering **86,** 283, (2009).
[2] V. Ioannou-Sougleridis, P. Dimitrakis, V. Em. Vamvakas , P. Normand,. C. Bonafos, S. Schamm, N. Cherkashin, G. Ben Assayag, M. Perego, M.Fanciulli, Microelectronics Engineering **84** 1986 (2007).
[3] V. A.Gritsenko, K. A. Nasyrov, Yu. N. Nikolov, A. L. Aseev, S. Y. Yoon, J.-W. Lee, E.-H. Lee, and C.W. Kim, Solid-State Electronics **47**, 1651, (2003).
[4] S. I. Shim, F. C. Yeh, X. W. Wang, T. P. Ma, IEEE Electron Device Letters 29, 512, (2008).
[5] C. L. Dezelah, J. Niinisto, K. Kukli, F. Munnik, J. Lu, M. Ritala, M. Leskela, and L.Niinisto, Chemical Vapor Deposition **14**, 358, (2008).
[6] K. Kukli, M. Ritala, T. Sajavaara, J. Keinonen, and M. Leskela, Chemical Vapor Deposition **8**, 199, (2002).

Mater. Res. Soc. Symp. Proc. Vol. 1250 © 2010 Materials Research Society   1250-G07-04

Low Temperature Growth of Silicon Structures for Application in Flash Memory Devices

Thomas A. Mih, Shashi Paul[1], Richard B. M. Cross
Emerging Technologies Research Centre, De Montfort University, LE1 9BH Leicester,
United Kingdom

ABSTRACT

An in-depth study of the structural and electrical properties of silicon (Si) films deposited by a novel low temperature technique at temperatures less than 400°C in a 13.56 MHz RF PECVD reactor is reported. The method is based on substrates having to undergo some initial preparatory steps (IPS) before the deposition of Si films in the PECVD chamber. The optical band gap of Si films deposited using this novel technique narrowed to 1.25 eV from 1.78 eV using the traditional a-Si:H deposition recipe. No annealing of any form was performed on the films to attain this band gap. Furthermore, photosensitivities for these films under various deposition conditions were of order 10^0 compared to 10^4 for a-Si:H films deposited under like conditions. Using metal-insulator-semiconductor devices, the Si films grown by this novel technique exhibit charge storage and memory behaviour unlike their amorphous counterparts. However, device endurance has been found to be inadequate, probably due to the presence of some contaminants - notably interstitial oxygen - which has been found elsewhere to have adverse effects on the electrical characteristics of Si films. If well harnessed, we suggest Si structures grown by this novel growth technique could be well-suited for flash memory applications, particularly 3-D flash which requires process temperatures to be less than 400 °C.

INTRODUCTION

The continuous down-scaling of flash memory cell layers is approaching a dead-end, where leakage currents will increase significantly and impact data retention. This challenge, coupled with the requirements of dielectric quality [1], may result in less integration and performance gains leading to flash performance and reliability being seriously degraded. Three-dimensional (3-D) cell architecture is one solution suggested to avert these problems and boost the performance of flash devices [2, 3]. It is attractive as it permits the integration of long-retention and high-density cells without compromising device reliability [4]. However, high temperature processing of memory layers is not ideal for 3-D stacked memory architecture, as it stresses device structures - especially at interfaces between different materials.

We have developed a novel methodology, which involves an initial substrate preparatory ritual, for growing high-quality silicon (Si) structures at $\leq$ 400 °C [5]. The advantage of this method compared with solid-phase crystallization (SPC) of a-Si:H is that it by-passes the long hours of annealing necessary for the SPC of a-Si:H. We have previously demonstrated the suitability of this growth technique for future 3-D flash memory technology [6]. In this study, we have used metal-insulator-semiconductor (MIS) devices, which mimic memory devices, to demonstrate the suitability of the low temperature-grown Si structures for application in flash memory devices. Capacitance-Voltage (C-V) studies were undertaken to understand the

---

[1] Corresponding author email: spaul@dmu.ac.uk

charging and retention time. An in-depth study of the structural, optical and electrical properties of thin-film Si structures grown by this novel technique is also be presented.

EXPERIMENT

In this study, p-type Si wafer (1-10Ω, Boron doped) and glass substrates were used and prepared as described in reference [6]. Si thin films were grown on the substrates from pure silane (SiH_4) as a precursor gas in a 13.56 MHz RF PECVD reactor. Growth conditions were varied in the temperature range 250 to 400 °C, a pressure range of 150 to 500 mtorr, an RF power range of 5 to 25 W, and a SiH_4 flow rate range of 10 to 50 sccm. The base pressure of the PECVD reactor before every deposition was within 3 to 10 mtorr.

Film thicknesses were obtained using an Alpha-Step 200 stylus profilometer for determining the growth rate of the films under different growth conditions. Ultraviolet-Visible (UV-Vis) spectroscopy was performed on the films on glass substrates using an Evolution 300 UV-Visible spectrophotometer. FTIR measurements were also carried out on films grown on p-type silicon substrates to obtain valuable information about film composition. In order to carry out electrical conductivity measurements of the Si films on glass substrates, gap-cells shadow masks were used to thermally-evaporate aluminum contacts on the films. The gap dimensions were 100, 250, 500 and 1000 µm respectively. Dark and photoconductivity measurements were then investigated in a light-tight box in the dark and under illumination with a 100mW/cm^2 Oriel solar simulator. Charge storage and memory capabilities of the films were investigated by C-V measurements using an HP 4192A and an HP 4082B on metal-insulator-semiconductor (MIS) devices, (schematic diagram shown in figure 1). The silicon nitride film was deposited in the PECVD reactor from SH_4, NH_3 and N_2 at flow rates of 40, 10 and 100 sccm respectively and at a temperature of 300°C, pressure 350 mtorr and RF power of 50W. The nitride film layer thickness measured using the ellipsometer was approximately 130 nm and the evaporated aluminum gate thickness was ≈ 100 nm.

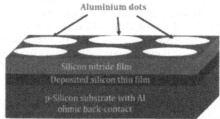

Figure 1 Metal-insulator-semiconductor device schematic diagram used to investigate the memory effect of thin Si films deposited on p-type silicon substrates with an aluminum back contact.

RESULTS AND DISCUSSION

Optical band gaps of the deposited films were estimated by using Tauc plots $(\alpha E)^{1/2}$ vs. E where E is the photon energy and α is the absorption coefficient, obtained from UV-Vis transmittance spectra. Films deposited on substrates without having undergone the initial preparatory step (non-IPS) showed band gaps in the range between 1.70 and 1.90 eV while those grown on IPS substrates exhibited band gaps between 1.25 to 1.70 eV. Figures 2a and 2b show the Tauc plots for the Si film deposited on a glass substrate for non-IPS (figure 2a) and IPS (figure 2b) sections of the same substrate. The film was deposited at 400 °C; 25W RF power and a chamber pressure of 200 mtorr at a SH$_4$ flow rate of 20 sccm. The dotted-broken lines are an extrapolation of the linear part of the curves to the photon energy axes at zero absorption. The intersection on this axis gives the band gap energy, 1.78 eV and 1.25 eV for non-IPS (figure 2a) and IPS (figure 2b) respectively. It was further observed that the band gap decreased with an increase in temperature, with films on IPS substrates showing a more significant change than the non-IPS films (figure 2c).

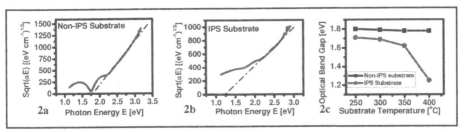

Figure 2. Tauc plots for films grown on (a) non-IPS and (b) IPS substrates at . Figure 2c is the band gap variation with deposition temperature at 200 mtorr pressure, 25 W RF power and 20 sccm silane flow rate.

The decrease in optical band gap was accompanied by a corresponding decrease in hydrogen content of the films grown on IPS substrates in comparison with those grown on non-IPS substrates. The hydrogen content of the films was estimated from $C_H = N_H/N_{Si}$ where N_{Si} is the total atomic density of the film ($5.3*10^{22}$ cm^{-3} for c-Si), N_H is the hydrogen concentration which is related to the integrated absorbance I by N_H = AI where I is given by [7]

$$I = \int \left(\frac{\alpha(\omega)}{\omega} \right) d\omega \qquad (1)$$

$\alpha(\omega)$ being the coefficient of absorption and ω the wave number in cm^{-1}. A Gaussian fit was used to estimate the integral in equation (2) over the absorption band centered on the Si-H wagging mode at 640 cm^{-1}. The proportionality constant A= $2.1*10^{19}$cm^{-2} for the Si-H wagging mode 640 cm^{-1}as obtained by Langford et al [7] was used. C_H was estimated to be ≈23.1 at.% for the Si films on the non-IPS substrate and ≈10.6 at.% for the IPS substrate for the films grown under the conditions described above. The C_H values were also found to decrease with an increase in SiH$_4$ flow rate both for IPS and non-IPS substrates between 10-30 sccm (graph not shown).

The presence of the peak at the vibration mode 1000 – 1100 cm^{-1} (Figure 3) in FTIR spectra revealed the presence of interstitial oxygen in all the films grown. Reports in the literature point to the interstitial oxygen being introduced during growth or during film exposure to air after growth as noted by Müllerová et al [8]. Hiraki [9] has noted an increase in absorbance around oxygen characteristic peaks with exposure time within the first 20 hours of exposure and suggests the oxygen inclusion in the films is a diffusion limited reaction. Furthermore, Torres et

11

al [10] also observed that the presence of oxygen has a profound adverse effect on the electrical conductivity of the film and causes the n-type behaviour of the material. The source and effects of this oxygen in our films are not exactly known and are being investigated.

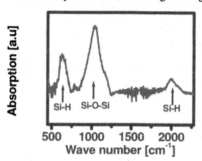

Figure 3 An FTIR absorption curve showing the Si-H wagging mode at 640 cm⁻¹ and the Si-O-Si asymmetric stretching vibration mode at 1070 cm⁻¹ and the Si-H stretching mode at 2000 cm⁻¹

The variation in dark- and photoconductivity with substrate temperature for Si films on non-IPS (Figure 4a) and IPS (Figure 4b) substrates reveal a stark difference between the films. The dark conductivity is at least 3 orders of magnitude lower for non-IPS compared with IPS substrate films. Furthermore, it was observed that the photo- and dark conductivity for IPS films was about the same order of magnitude and increase with temperature. The photosensitivity which is the ratio of photo- to dark conductivity (Figure 4c) is within 10^2-10^4 for non-IPS substrate films and 10^0 for IPS substrate films.

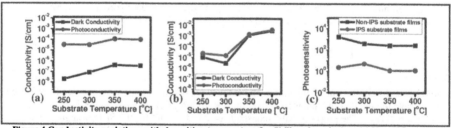

Figure 4 Conductivity variations with deposition temperature for Si films deposited on (a) Non-IPS and (b) IPS substrates. Figure 4c is the photosensitivity variation with deposition temperature.

C-V characteristics show very small hysteresis for MIS devices containing only silicon nitride on non-PIS substrates (Figure. 5a), as well as for silicon nitride and silicon film on non-IPS part of the substrate (Figure. 5b). A large hysteresis of ≈4V is obtained for Si films on the IPS part of substrate (Figure. 5c). The hystereses were anticlockwise for sweep voltages from positive to negative and back to positive. The small hysteresis in 5a and 5b could be attributed to defects and interface traps in the silicon nitride dielectric, meanwhile that in 5c is largely due to stored charge in the poly-Si film grown on the IPS substrate injected from the substrate as voltages were swept from +4V to -8 or -10V and back to +4V. Previous results using polystyrene and polyvinyl acetate as a dielectric did not show any hysteresis in devices deposited on non-IPS substrates [6].

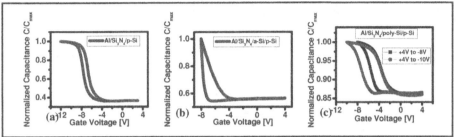

Figure 5 C-V characteristics for MIS devices (a) Al/Si₃N₄/p-Si, (b) Al/Si₃N₄/non-IPS Si Film/p-Si, (c) Al/Si₃N₄/IPS Si Film/p-Si

The memory behaviour was investigated for the Si films on the non-IPS and on the IPS substrates. It was found that approximately the same value of capacitance was obtained during a read operation after each write and erase steps (Figure 6a); but for the IPS substrate films, two capacitance states could be identified which remained fairly constant during the read operation after every write and erase operation (Figure 6b). These states are identical to the logic levels of 0 and 1 or high and low in proper memory devices. Non-volatility was also observed, even though the write and erase curves soon crossed after a few write/read/erase cycles (graph not shown).

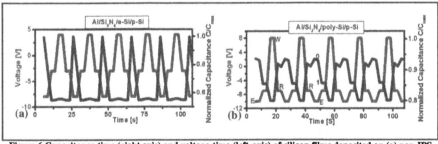

Figure 6 Capacitance-time (right axis) and voltage-time (left-axis) of silicon films deposited on (a) non-IPS substrates and (b) IPS substrates. In (a) the device was written at +4V and erased at -8V then read at -3V. In (b) the write voltage was 8V, Erase -10V and read voltage was -7V

CONCLUSION

A novel low temperature deposition technique of Si structures has been investigated. An optical band gap of 1.25 eV of the silicon films grown by the novel method has been obtained compared to 1.78 eV for films deposited under the same conditions and on the same substrate - a part of which has been specially treated. The films deposited by the new method show very

13

distinct electrical conductivities and photosensitivity from those grown by traditional PECVD methods. The photo- to dark conductivity ratio was $\approx 10^0$ for the novel method and $\approx 10^3$ for the traditional method. C-V characteristics also exhibited charge storage in films deposited using the novel methodology. Furthermore, capacitance-time curves showed memory behaviour only for films grown by this new method and not for films deposited by the traditional technique. It is worth noting also that no annealing was done to any of the films reported in this work. The films however exhibited the presence of oxygen contaminant that we think has an adverse effect on the memory characteristics of the film. The effects will be investigated and reported in due course. All the properties of the films investigated so far are suggestive of the fact that these films are polycrystalline in nature. However, the exact nature of the structures grown is being further investigated using XRD and other material structure characterization techniques. The material is highly promising for flash memory applications especially in 3-D flash memory requiring process temperatures to be less than 400°C.

REFERENCES

1. Greg Atwood; Future directions and challenges for ETox Flash memory scaling, IEEE Transactions on Device Materials Reliability **Volume 4, (2004),** pp. 3001-305
2. D. Park, K. Kim, B.I. Ryu; "3-D nano-CMOS transistors to overcome scaling limits" in *Proc. Solid-state and Integr. Circuits Technol.* **(2004)**, 1, 34-35
3. S. Koliopoulou, P. Dimitrakis, D. Goustouridis, P. Normand, C. Pearson, M.C. Petty, H. Radamson and D. Tsoukalas, *Microelectronic Engineering*, 83, **(2006)**, 1563
4. R. Tsuchiya, M. Izawa, S. Kimura, "Prospects of Si semiconductor devices and manufacturing technologies in nanometer era", *Hitachi Review* **(2006)**, 55(2), 46-55
5. S. Paul (This work will be communicated to nature; shortly).
6. T.A. Mih, R.B. Cross, S. Paul; "A novel method for the growth of low temperature polycrystalline silicon for 3-D flash memory" in *Materials and Technologies for 3-D Integration*, edited by F. Roozeboom, C. Bower, P. Garrou, M. Koyanagi, P. Ramm (Mater. Res. Soc. Symp. Proc. **Volume 1112,** Warrendale, PA, **2009)** pp1112, 265-269
7. A. A. Langfold, M. L. Fleet, B. P. Nelson and M. Marley; "Infrared absorption strength and hydrogen content of hydrogenated amorphous silicon" Phys. Rev. B, **(1992),** 45, 13367
8. J. Müllerová, S. Jurečka and P. Šutta; "Optical characterization of polysilicon thin films for solar applications", *Solar Energy* **(2006),** 80, 667-674
9. A. Hiraki; "Impurity effects" in Amorphous semiconductor Technologies and Devices, edited by Y. Yamakawa, (Japan Annual Reviews in Electronics, Computers and Telecommunication **Volume 16,** North Holland **1984**), pp.134-148
10. P. Torres, J. Meier, R. Flu¨ ckiger, U. Kroll, J. A. Anna Selvan, H. Keppner, and A. Shah; "Device grade microcrystalline silicon owing to reduced oxygen contamination", *Appl. Phys. Lett.,* **(1996),** 69(10), 1373-1375

Nano-Crystal Memory

Mater. Res. Soc. Symp. Proc. Vol. 1250 © 2010 Materials Research Society   1250-G01-10

OVERVIEW OF ADVANCED 3D CHARGE-TRAPPING FLASH MEMORY DEVICES

Hang-Ting Lue, Kuang-Yeu Hsieh, and Chih-Yuan Lu

Macronix International Co., Ltd.,
16 Li-Hsin Road, Hsinchu 300, Taiwan, R.O.C.
e-mail: htlue@mxic.com.tw

ABSTRACT

Although conventional floating gate (FG) Flash memory has already gone into the sub-30 nm node, the technology challenges are formidable beyond 20nm. The fundamental challenges include FG interference, few-electron storage caused statistical fluctuation, poor short-channel effect, WL-WL breakdown, poor reliability, and edge effect sensitivity. Although charge-trapping (CT) devices have been proposed very early and studied for many years, these devices have not prevailed over FG Flash in the > 30nm node. However, beyond 20nm the advantage of CT devices may become more significant. Especially, due to the simpler structure and no need for charge storage isolation, CT is much more desirable than FG in 3D stackable Flash memory. Optimistically, 3D CT Flash memory may allow the Moore's law to continue for at least another decade. In this paper, we review the operation principles of CT devices and several variations such as MANOS and BE-SONOS. We will then discuss 3D memory architectures including the bit-cost scalable approach. Technology challenges and the poly-silicon thin film transistor (TFT) issues will be addressed in detail.

Introduction Charge-Trapping (CT) NAND Flash

Figure 1 briefly compares the FG and CT NAND. CT NAND has several advantages over FG: It can be integrated in a simpler planar structure, and the discrete trap charge storage can suppress interference and stress induced leakage current (SILC) issues. For 3D memory, it is not necessary to cut through the ONO charge-trapping layer, and this greatly simplifies the 3D memory process and enables the ultra low cost memory. Figure 2 illustrates several examples in previous works [1-3].

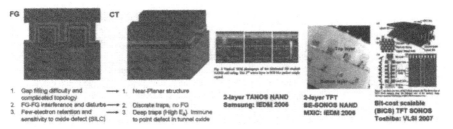

Fig. 1 Comparison of FG NAND and CT NAND.   Fig. 2 Several examples of 3D CT NAND Flash devices. [1-3]

17

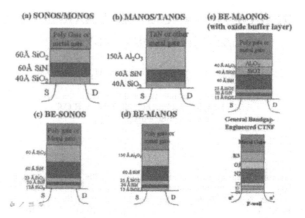

Fig. 3 Comparison of FG NAND and CT NAND.

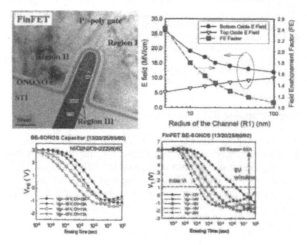

Fig. 4 FinFET BE-SONOS and Field enhancement effect.

Figure 3 briefly illustrates several important candidates for charge-trapping devices [4-8]. The conventional SONOS/MONOS can not find a suitable operation window due to the difficult trade off between the erase and retention. Two approaches were extensively proposed. The first one is to use high-K top dielectric and a high work function metal gate (MANOS) to suppress the gate injection and erase saturation [6]. However, the erase speed is still very slow because the major erase mechanism comes from electron de-trapping out of nitride. To expel the deeply trapped electrons out of nitride is very inefficient, and it shows a gradually slower erase speed due to the electron trapped energy "spectrum blue shift" [9]. Moreover, the very high field during erase also damages the tunnel oxide reliability. To further improve the electron detrapping speed one must detune the nitride to a "shallowly trapping" material such as silicon-rich nitride. However, again the trade-off between the erase speed and retention does not allow a reliable device.

Another important approach is to use the barrier engineered (BE) tunneling barrier [4] to offer fast hole injection at high field, while eliminate the direct tunneling leakage at low field and preserve good retention. Using hole injection to erase the CT device is very important to avoid the difficult trade off between data retention and erase speed - since shallowly trapping SiN material is not used electrons can be deeply trapped, thus preserving the original merit of CT devices. The BE CT device can also be combined with the high-K top dielectric and metal gate to further suppress the erase saturation and open a larger memory window (BE-MANOS). We also find that using a thin high-K top dielectric together with a buffer oxide in between SiN is important to provide good reliability (BE-MAONOS). In addition, a thinner high-K is important to minimize the bulk trapped charge in high-K material and suppress the Vt instability such as fast retention loss and transient response [10].

In addition to the thin film engineering, CT device is also sensitive to geometry and edge effect [11]. In Fig. 4, when CT device is integrated in a rounded-top FinFET or nano wire structure with a small radius of curvature, the field enhancement (FE) effect takes place [12], resulting in dramatically improved programming/erasing speed. Although this FE effect can create larger memory window and faster speed, the sensitive dependence on the geometry makes the device critical to the geometry and the Vt distribution is wider. Moreover, the large FE also makes both read and program disturb worse. It should be carefully managed for a reliable NAND Flash product.

Chip-level Performance of BE-SONOS NAND Flash

The success of barrier engineered CT device has been proven. Figure 5 shows the typical P/E memory window of BE-SONOS NAND test chip [13] fabricated in a 75nm bulk Si technology. Figure 6 shows that BE-SONOS also possesses excellent retention under P/E cycling and high-temperature baking tests. One very important merit is that during P/E cycling and retention, there is no tail bit. This is a very important advantage of CT devices. Since charges are independently stored in discrete traps, CT device is naturally immune to the SILC issue common in the FG technology. We also found that the major endurance degradation comes from the tunnel oxide interface state generation (Dit), and can be minimized by using a nitrided O1 to strengthen the Si/O1 interface quality.

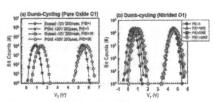

Fig. 5 (a) Dumb PE cycling (without any P/E verify) of BE-SONOS with pure oxide O1. (b) Dumb cycling of BE-SONOS with nitrided O1. Erase degradation is improved by nitrided O1. For both devices, dumb-cycling do not create a single tail bit. Edge WL's (WL0 and WL31) are excluded to avoid processing differences.

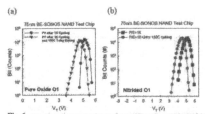

Fig. 6 Retention of programmed state after 1K P/E cycling and 150C high-temperature baking test for one block (128kb). (a) Pure oxide O1 (by ISSG process). (b) Nitrided O1. Although charge loss is observed after P/E cycling, no single tail bit is observed and the distribution is normal.

Compared to FG NAND at the same technology, BE-SONOS NAND has faster programming speed, narrower programmed Vt distribution, and no tail bit issues. However, the erase saturation is higher and this restricts BE-SONOS NAND's erase speed. Engineering solutions such as tuning the O1 barrier and using higher-K blocking oxide are being investigated.

TFT CT Devices Characteristics

Recently, the 3D NAND Flash becomes an exciting approach to extend the NAND Flash roadmap beyond 20 nm node [1-3, 14-18]. The most important element of 3D memory is the thin-film transistor (TFT) CT device, as shown in Fig. 7. In Fig. 8, we have found that as dimensions scale to sub-30nm, TFT memory has excellent performance [14]. The physical reason is that most sub-30nm TFT devices contain no grain boundary and are essentially single grain devices. As a result, the DC characteristics such as read current and subthreshold slope can approach the bulk Si devices.

However, when we investigate the distribution of many devices, certain tail distribution is observed, as shown in Fig. 9. The tail bits were found to have poor initial DC performance. Fortunately, the poor initial DC characteristic does not affect the P/E memory window since FN tunneling is independent of the grain boundary effect.

The tail distribution of TFT memory should be improved by further poly silicon engineering. Moreover, the NAND Flash also allows many error correction techniques so that tail bits issues of TFT memory may be overcome.

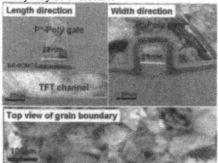

Fig. 7 A sub-30nm TFT BE-SONOS NAND Flash device [14].

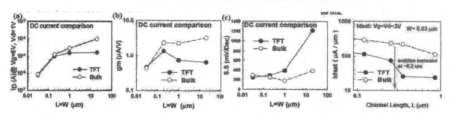

Fig. 8 Comparison of DC characteristics of sub-30nm TFT and bulk Si BE-SONOS NAND Flash device [14].

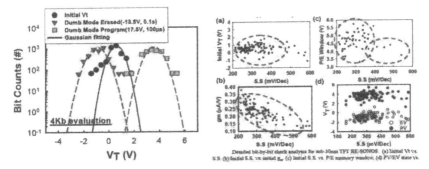

Fig. 9 P/E distribution and bit-by-bit tracking of the tail bits [14].

3D NAND Flash Architecture Using CT Devices

The TFT CT devices can be designed in various 3D array architectures. Figure 10 briefly compares several important candidates.

	P-BiCS	TCAT	VSAT	VG
String	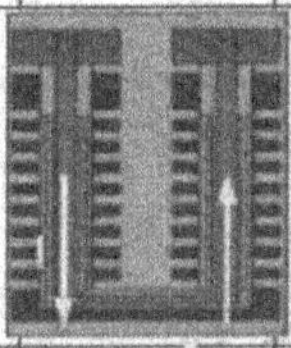			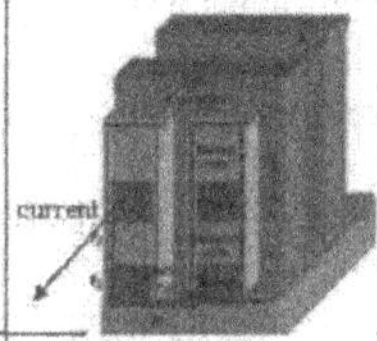 current
Cell Shape				
Cell Size in X, Y direction	6F² (3F*2F)	6F² (3F*2F)	6F²(3F*2F)	4F² (2F*2F)
Gate Process	Gate first	Gate last	Gate First	Gate Last
Current Flow direction	U-turn	Vertical	Multi-U-turn	Horizontal
Device Structure	GAA	GAA	Planar	Double Gate
Possible minimal F	~50 nm	~50 nm	~50 nm	~2X nm

Fig. 10 Comparison of various 3D NAND Flash array architectures [1, 15- 18].

21

The most important concept of 3D memory is the so-called bit-cost scalable (BiCS) approach [1]. Only one critical contact drill hole is used to carry out multi-layer stack of TFT memory, thus as layer number increases, the bit cost continuously decreases. Since the original BiCS structure has tunnel oxide damage issue, a pipe-shaped P-BiCS [15] was proposed to solve this issue. Latter, TCAT [16] was proposed to provide a gate last process that enables the use of metal gate. VSAT [17] and Vertical gate (VG) [18] are others approaches which do not use a drilled hole but instead line patterning processes.

It should be mentioned that all 3D NAND Flash architectures have lateral (X,Y) scaling limitation. First, the drill hole process is naturally less scalable than the conventional line patterning process due to the lithography limit. Second, both P-BiCS and TCAT must use additional WL cut process in the X direction, which limits the X-direction pitch. Third, the 3D memory architecture must fill in the trench with 2-side charge-trapping ONO layer, which limit the X-direction space.

Comparing the above four structures, we think that VG structure may have the best X, Y pitch scalability and may extend to ~2X nm node with multi-layer stack.

Another important issue of 3D NAND Flash is that they all have floating body structure, where the conventional substrate erase can not be applied. Therefore, for 3D memory, different erase methods such as utilizing the hole generated by the gate induced drain leakage current to charge-up the channel potential for FN erase is necessary.

The 3D memory is difficult to have conventional n^+ junction in between wordlines. Therefore, a so-called "junction-free" NAND [19, 20] is necessary to offer the 3D NAND Flash integration.

Finally, to offer a high-performance and reliable device the understanding and engineering design of CT device is the most central part. Using the barrier engineered (BE) CT device is the most important way to provide both excellent performance and reliability.

Summary

3D charge-trapping (CT) NAND Flash offers a promising path for NAND Flash scaling. Although it is difficult for CT to prevail the FG NAND at >20nm node, CT NAND is naturally suitable for 3D NAND Flash structures and enable the continuous scaling of NAND Flash.

ACKNOWLEDGMENTS

The authors would like to thank many colleagues in Emerging Central Lab. (ECL) and Technology Development (TD) center for their supporting in BE-SONOS NAND Flash.

REFERENCES
[1] H. Tanaka, et al, VLSI Symposia, pp. 14-15, 2007. [2] E. K. Lai, et al, IEDM, pp. 41-44, 2006. [3] S. M. Jung, et al, IEDM, pp. 37-40, 2006. [4] H. T. Lue, et al, IEDM, pp. 547-550, 2005. [5] H. T. Lue, et al, IRPS pp. 874-882, 2009. [6] C.H. Lee, et al., IEDM Tech. Dig., pp. 26.5.1 - 26.5.4, 2003. [7] S.C. Lai, et al., IEEE NVSMW, pp. 88-89, 2007. [8] S.C. Lai, et al., IEEE NVSMW, pp. 101-102, 2008. [9] H.T. Lue, et al. IRPS, pp. 168-174, 2005. [10] S. C. Lai, et al, IMW, 2010, in publication. [11] H. T. Lue, et al, IEDM, pp. 839-842, 2009. [12] T. H. Hsu, et al, IEEE T-ED, pp. 1235-1242, 2009. [13] H. T. Lue, et al, IRPS, 2010, in publication. [14] T. H. Hsu, et al, IEDM, pp. 629-632, 2009. [15] R. Katsumata, et al, VLSI Symposia, pp. 136-137, 2009. [16] J. Jang, et al, VLSI Symposia, pp. 192-193, 2009. [17] J. Kim, et al, VLSI Symposia, pp. 186-187, 2009. [18] W. Kim, et al, VLSI Symposia, pp. 188-189, 2009. [19] H. T. Lue, et al, VLSI Symposia, pp. 140-141, 2008. [20] H. T. Lue, et al, VLSI Symposia, pp. 224-225, 2009.

Mater. Res. Soc. Symp. Proc. Vol. 1250 © 2010 Materials Research Society   1250-G01-02

Annealing effects on Si nanocrystal nonvolatile memories

Panagiotis Dimitrakis[1], C. Bonafos[2], S. Schamm-Chardon[2], G. Ben Assayag[2], P. Normand[1]
[1]Institute of Microelectronics, NCSR "Demokritos", 15310 Aghia Paraskevi, Greece
[2]CEMES-CNRS, Université de Toulouse, nMat group, BP 94347, 31055 Toulouse cedex 4, France

ABSTRACT

The effect of thermal treatments in oxidizing ambient on the structural and electrical properties of low-energy Si-implanted thin SiO_2 layers which previously subjected or not to high temperature annealing in inert ambient was investigated. Based on TEM examination, charge trapping evaluation and FN conduction analysis of the resulting Si-NC SiO_2 matrices, a model taking into account the healing of excess silicon atoms introduced by implantation and the generation of Si interstitials by oxidation above and below the viscoelastic temperature of SiO_2 is proposed.

INTRODUCTION

The nanocrystal (NC) nonvolatile memory (NVM) devices are potential candidates for replacing the conventional polysilicon floating gate memories at technology nodes below 45nm [1]. Nanocrystals of a large number of materials have been investigated in order to optimize the operation conditions, the performance and the reliability of NC-NVMs. A significant amount of research has been carried out utilizing semiconducting NCs, like Si and Ge. The Si-NCs seem to meet most of the processing integration requirements for future manufacturing in a CMOS environment [2]. A promising technique to realize self-organized Si-NCs into thin tunnel oxide layers is the ultra-low-energy ion-beam-synthesis (ULE-IBS) that basically comprises the steps of ion implantation at an energy ≤1keV and subsequent annealing [3,4]. ULE-IBS of Si-NCs is of special interest due to its inherent compatibility with CMOS processing. Attractive device results have been presented using such a technique demonstrating low-voltage operation at reasonable speeds, endurance to repeated write/erase cycles and 10-year data retention. [5]. Different experiments revealed that in addition to the ion-implantation parameters [9], the post-implantation annealing conditions play a key role on the operation and performance of Si-NC-NVMs [6-8]. Among the different annealing atmospheres that have been tested, annealing in presence of a small percentage of oxygen has been proved to be a useful method to control the size and the density of Si-NCs as well as to improve the integrity of the implanted oxide. Although, the effect of oxidation on Si-NCs has been studied and modeled thoroughly [10], it is not well understood how oxygen affects the memory performance.

In this paper, we present structural and electrical studies conducted on oxygen-annealed Si-implanted thin SiO_2 layers, which previously suffered or not high temperature annealing in inert ambient. More specifically, we investigate the effect of annealing time in oxidizing ambient on the SiO_2 regions located above and below the NCs layer, which are commonly referred to as the control (CO) and tunnel (TO) oxides respectively. The existence of Si residuals in both layers significantly affect their trapping and conduction properties and hence the performance of the final structures for memory applications.

23

EXPERIMENTAL AND METHODS

High quality 7nm SiO_2 layers were thermally grown on 200mm p-Si wafers in a manufacturing environment and subsequently implanted with 1keV Si ions at a dose of 1×10^{16} or 2×10^{16} Si^+/cm^2. Then, the wafers were cut into pieces and different thermal treatments were performed. The high-dose implanted samples were annealed at 950°C in nitrogen-diluted-oxygen (1.5% O_2 per volume) for durations in the 5-120 min range (XP samples, see Table I) while a two-step annealing process made of a first annealing at 1050°C in N_2 for 30 min and a second annealing at 900°C in $N_2/1.5\%$ O_2 for various durations was applied to the low-dose implanted samples (II samples, see Table II). Previous experiments [3,5-8] revealed that the above one-step annealing process is the most suitable for highly Si implanted SiO_2 layers in terms of memory performance. The two-step annealing process leads for the high dose samples to connected NCs arrays during the high temperature (1050°C) treatment. In that case, a subsequent thermal treatment at a lower temperature in N_2/O_2 does not allow neither formation of well-defined and separated NC arrays nor efficient healing of the TO. On the contrary, the low-dose implanted samples take advantage of the enhanced Si ripening into NCs that occurs during the first high-temperature annealing regime leading to the formation of uniform and well-separated Si NCs with a mean diameter of 2.9 ± 0.3 nm [10]. In that case, the following thermal treatment at 900°C in N_2/O_2, which is below the viscoelastic temperature (~950°C) of SiO_2 [10], has limited effects on the NC size due to self-limited oxidation and increased efficiency in TO healing. Extensive TEM studies were performed in order to estimate the total thickness of the processed SiO_2 layers as well as the CO and TO thicknesses.

Table I. One-step Annealing conditions for high-dose samples. The initial implanted SiO_2 layer and the implantation energy were 7nm and 1KeV respectively.

S/N	Implantation Dose (Si^+cm^{-2})	Annealing conditions				Sample ID
		Temperature (°C)	$\dfrac{[O_2]}{[N_2]+[O_2]}$ %	Time (min)		Sample ID
1	0	950	1.5	30		XP7R9
2	2×10^{16}	950	1.5	5		XP7313
3	2×10^{16}	950	1.5	15		XP7314
4	2×10^{16}	950	1.5	30		XP7315
5	2×10^{16}	950	1.5	120		XP7316

Standard Al-gate MOS capacitors were fabricated for capacitance (C-V) and current voltage (I-V) measurements of the annealed structures. C-V measurements allow for the extraction of memory windows obtained by bi-directional voltage sweeps. The I-V measurements were carried out under application of positive and negative voltages and analyzed at high electric field regime (Fowler-Nordheim regime, FN) for the extraction of FN barriers for both bias polarities. In these calculations an electron effective mass in SiO_2 of $0.5m_0$, where m_0 is the free electron mass, was assumed. Furthermore, I-V characteristics were used to calculate the resistivity ρ of the implanted SiO_2 layers after each annealing process. According to Card et al [11] the resistivity ρ of the SiO_2 dielectrics as a function of the applied electric field E in the FN regime is following a simple exponential relation $\rho = \rho_0 \exp(-\alpha E)$, where ρ_0 and α are the characteristic parameters extracted from the linear part of the resistivity plot $\ln(\rho)$-E.

Table II. Two-step annealing conditions for low-dose samples. The initial implanted SiO$_2$ layer and the implantation energy were 7nm and 1KeV respectively.

S/N	Implantation Dose (Si^+cm^{-2})	Annealing conditions	Sample ID
1	0	1050°C, N$_2$, 30min/ 900°C, N$_2$, 30min	IIR
2	1×10^{16}	1050°C, N$_2$, 30min/ 900°C, N$_2$, 30min	II20
3	1×10^{16}	1050°C, N$_2$, 30min/ 900°C, [O$_2$]=1.5%, 5min	II21
4	1×10^{16}	1050°C, N$_2$, 30min/ 900°C, [O$_2$]=1.5%, 30min	II22
5	1×10^{16}	1050°C, N$_2$, 30min/ 900°C, [O$_2$]=1.5%, 60min	II23
6	1×10^{16}	1050°C, N$_2$, 30min/ 900°C, [O$_2$]=1.5%, 120min	II24
7	1×10^{16}	1050°C, N$_2$, 30min/ 900°C, [O$_2$]=1.5%, 240min	II25

RESULTS AND DISCUSSION

TEM studies

The overall thickness of the SiO$_2$ layers and the thicknesses of the TO and CO layers as extracted from cross-section TEM examination for the XP and II samples are reported in Table III and IV respectively. Obviously, following the single step annealing process the thicknesses of the tunnel and control oxides increase with the annealing duration indicating that oxygen diffusion and further oxidation of part of the implanted Si atoms are not time limited and take place simultaneously with the Si-NCs formation. For samples treated with the two-step annealing process, the density and diameter of NCs vary in the range $1.7\times10^{12} - 5\times10^{12}$ cm^{-2} and 1.1-2nm respectively. Significant changes in thickness of the TO layer are detected after 900°C-N$_2$/O$_2$ annealing times up to 60 min. The origin of this enhanced oxidation effect of the TO compared to the CO layer is not obvious and electrical investigations are necessary in order to draw accurate conclusions.

Electrical characterization studies

Figures 1(a) and 1(b) present the memory window in terms of flat-band voltage V$_{FB}$ shift following different voltage sweeps in C-V measurements. In the case of the single-step annealing process up to 30min, voltage sweeps from inversion to accumulation cause electron injection from Si substrate and subsequent trapping into Si-NCs. After 5min annealing a network of interacting elongated Si-NCs is formed causing lateral loss of charges [8]. This observation in combination with the low consumption of excess Si atoms and the poor recovery of implantation-induced defects explain both the charging effects observed at very low voltages (< 2V) and the poor charge storage properties of this particular sample. As the annealing time increases the Si-NCs are oxidized and become mutually isolated [8] allowing for the faster diffusion of oxygen atoms towards the TO/Si interface. The early charging (>2V) detected in the 15 and 30min annealing cases reveals an electron injection process which may be attributed to both (i) the small energy barrier Φ_{BS} the electrons face at the Si/TO interface and (ii) the presence of near interface oxide traps due to the Si-rich nature of TO. Voltage sweeps from accumulation to inversion cause hole injection from the substrate to Si-NCs. According to TEM results the TO thickness is always thicker than the CO. However, electron tunneling from the gate was never observed (i.e. bi-directional C-V characteristics exhibit counter-clockwise hysteresis) suggesting that the CO is less Si/defect-rich compared to the TO layer and thereby, exhibits a higher energy barrier Φ_{BM} for electrons. Prolonged oxidation time (i.e., 120 min) reduces the memory window and at very high voltage sweeps (±8V) electron injection from the gate occurs. These are indications that there is a small concentration of tiny Si-NCs which act as charge trapping centers at high voltages. Due to the faster oxidation of the TO and its resulting enhanced dielectric properties, the large difference in thickness (1.5nm, Table III) between the TO and CO layers does not allow the observation of charge storage initiated from substrate carriers.

Table III. XTEM measurement results obtained from XP samples.

Sample ID	Total thickness of implanted SiO$_2$ (nm)	Thickness of TO (nm)	Thickness of CO (nm)
XP7313	12.6	5.8	4.5
XP7315	13.2	6	5.1
XP7316	15.2	7.5	6

Table IV. XTEM measurement results obtained from II samples.

Sample ID	Total thickness of implanted SiO$_2$ (nm)	Thickness of TO (nm)	Thickness of CO (nm)
II20	10.4	4.4	3.8
II21	10.2	4.4	3.9
II22	11.3	5.3	4.2
II23	11.3	5.3	4.1
II24	12.2	–	–
II25	13.2	–	–

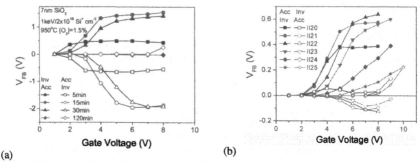

(a)   (b)

Figure 1. Memory windows obtained from samples described in (a) Table I and (b) Table II.

On the contrary, samples treated following the two-step annealing process exhibit clockwise hysteresis resulting to positive (negative) V_{FB} shift when the voltage is swept from accumulation to inversion (inversion to accumulation) as clearly presented in Fig.1(b). Taking into account the TEM results shown in Table IV, we conclude that this clockwise hysteresis effect is due to the lower CO thickness compared to the TO and that the insulating properties of the TO following the two-step annealing process are enhanced in comparison to those of the TO layer of the single-step annealed samples. Meaning that the TO has improved values of Φ_{BS}, is less Si-rich and has a lower trap density.

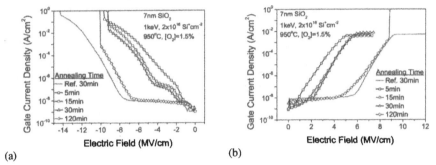

(a)   (b)

Figure 2. J-E characteristics for samples described in Table I under (a) negative and (b) positive applied electric fields.

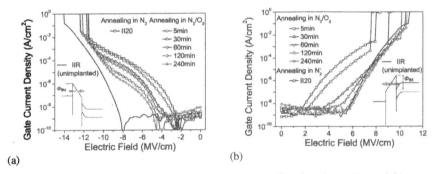

(a)   (b)

Figure 3. J-E characteristics for samples described in Table II under (a) negative and (b) positive applied electric fields.

Figures 2(a) and (b) show the current density, J, as a function of the applied electric field, E, for the one-step annealed samples. The improvement in current conduction with annealing time is obvious and can be attributed to the increase quality of the oxides resulting from the consumption of excess silicon and recovery of implantation-induced defects. Furthermore, calculations of the Φ_{BS} and Φ_{BM} barriers from the related FN-plots (not shown here) revealed that the barrier heights progressively increase (almost linearly) with the annealing time. This is inline with the TEM measurements presented above. Nevertheless, conductivity calculations (not shown here) suggest that ρ_0 is more than one order of magnitude higher for the CO compared to the TO except for the non-implanted and 5 min annealed samples. Finally, it is clear that the prolonged annealing time of 120 min leads to CO and TO layers with conduction properties very similar to that of the non-implanted sample indicating that the implanted layers are almost recovered.

Figures 3(a) and (b) present the J-E plots of the samples treated by the two-step annealing process. The improvement in current conduction with annealing time is faster for applied positive electric fields. Fig.3(b) indicates that for oxidation times longer than 5min the J-E characteristics are very similar, with small deviations below 6MV/cm. On the other hand, such a picture does not apply for negative electric fields. These results denote that the insulating properties of the TO recover faster than the CO ones. This is also revealed by the Φ_{BS} and Φ_{BM} barrier values extracted from the related FN-plots and resistivity calculations (not shown here).

We can thus propose the following scenario: At low temperatures (below 950°C, the SiO_2 viscoelastic temperature) there is a healing of the control and injection oxides due to the oxidation of the excess Si generated by implantation that remains after the step of NC formation at high temperature. For high-dose samples, the insulating properties of CO are improved compared to those of TO due to the blocking of the diffusion of oxidizing species by the connected network of the elongated Si-clusters at the first stages of the annealing. Healing of the TO layer is more efficient than in the case of the single-step annealing process. This may be attributed to the enhanced diffusion of the oxidizing species towards the substrate interface due the higher temperature of Si-NC formation and the lower implanted dose.

REFERENCES

1. *International Technology Roadmap for Semiconductors, ITRS*, 2007, http://www.itrs.net.
2. P.Dimitrakis, P.Normand and D.Tsoukalas, *"Silicon Nanocrystal Memories"*, pp.211-241 in "Silicon Nanopthonics", ed. L.Khriachtchev, World Scientific Publishing, Singapore, 2008.
3. P.Dimitrakis and P. Normand, D5.1, Mater. Res. Soc. Symp. Proc. 830, eds. A. Claverie, D. Tsoukalas, T-J. King, and J.M. Slaughter, Warrendale, PA , 2005
4. V. Beyer, J. von Borany, *Ion-beam Synthesis of Nanocrystals for Multidot Memory Structures*, in *Materials for information Technology*, eds. E.Zschech, C.Whelan, T.Mikolajick, Springer-Verlag London 2005
5. P. Dimitrakis, E. Kapetanakis, D. Tsoukalas, D. Skarlatos, C. Bonafos, G. Ben Asssayag, A. Claverie, M. Perego, M. Fanciulli, V. Soncini, R. Sotgiu, A. Agarwal, M. Ameen, P. Normand, Solid-State Electronics, 48, 1511-1517, (2004)
6. P. Dimitrakis, E. Kapetanakis, P. Normand, D. Skarlatos, D. Tsoukalas, K. Beltsios, A. Claverie, G. Benassayag, C. Bonafos, D. Chassaing, V. Soncini, Mater. Scien. Engineer. B, 101, 14-18 (2003)
7. P. Normand, E. Kapetanakis, P. Dimitrakis, D. Tsoukalas, K. Beltsios, N. Cherkashin, C. Bonafos, G. Benassayag, H. Coffin, A. Claverie, V. Soncini, A. Agarwal, M. Ameen, Appl. Phys. Lett., 83, 168-170 (2003)
8. C. Bonafos, H. Coffin, S. Schamm, N. Cherkashin, G. Ben Assayag, P. Dimitrakis, P. Normand, M. Carrada, V. Paillard and A. Claverie, Solid-State Electronics, 49, 1734-1744 (2005)
9. D. Tsoukalas, P. Dimitrakis, S. Koliopoulou and P. Normand, Mater. Scien. Engineer. B 124-125, 93-101 (2005)
10. H. Coffin, C. Bonafos, S. Schamm, N. Cherkashin, G. Ben Assayag, A. Claverie, M. Respaud, P. Dimitrakis, and P. Normand, J. Appl. Phys. 99, 044302 (2006)
11. H.C.Card and M.I.Elmasry, Solid-state Electronics 19, 863-870 (1976)

Mater. Res. Soc. Symp. Proc. Vol. 1250 © 2010 Materials Research Society   1250-G01-04

Theoretical Characterization of a Nanocrystal Layer for Nonvolatile Memory Applications.

Yann Leroy, Dumitru Armeanu and Anne-Sophie Cordan

Institut d'Électronique du Solide et des Systèmes – InESS (UMR 7163, CNRS-UdS), ENSPS, Bd Sébastien Brant, BP 10413, Illkirch, 67412, FRANCE.

ABSTRACT

On the road to miniaturization, nanocrystal layers are promising as floating gate in nonvolatile flash memories. Although much experimental work has been devoted to the study of these new memory devices, only few theoretical models exist to help the experimentalists to understand the physical phenomena encountered and explain the behavior of the device.

We have developed a model based on the geometrical and physical properties of the elementary structure of a nanocrystal flash memory, *i.e.* one nanocrystal embedded in an oxide between the channel and the gate electrodes. To obtain a fine analysis of the observed phenomena, several specific hypotheses have been taken into account. Concerning the channel, the contribution of the subbands is explicitly included. In the case of an electrode with a quasi-continuum of energy levels, we replace the continuum by equivalent sets of 2D subbands in order to be able to isolate the energy range that really contributes to the charging/discharging of the nanocrystal. The properties of the materials (bulk band structure, dielectric permittivity, ...) can be easily set as well as the geometrical specifications of the elementary structure (nanocrystal radius, tunnel and control oxide thicknesses, ...).

The behavior of a layer of nanocrystals is described according to a statistical approach starting from single nanocrystal results. This method allows us to take into account the fluctuations of geometrical parameters. Thus we are able to simulate various types of materials for the nanocrystals (Si, Ge, ...), the oxide layer (SiO_2, HfO_2, ...) and the electrodes, for both a single nanocrystal and layers of nanocrystals.

INTRODUCTION

On the road to miniaturization, nanocrystal (NC) layers are promising as floating gate in nonvolatile flash memories. Although much research has been carried out in this field [1–5], only some theoretical realistic models (other than the 1D approaches) exist to help the experimentalists to understand the physical phenomena encountered and explain the behavior of the device [6–10]. Among those models, few are able to account for disordered effects coming from geometrical fluctuations that originate from elaboration process [8, 10]. Indeed the boundary conditions commonly used to limit the spatial extension are equivalent to ordered layers of identical NCs.

To circumvent this limitation we have developed an approach centered on the description of a single isolated NC embedded in an insulator between two electron reservoirs (a p-doped Si channel and a metallic Au gate) [10,11]. The figure 1 shows the schematic representation of the NC floating gate memory under consideration (fig. 1a) and the cross-section of the isolated NC, together with the characteristic geometrical parameters (fig. 1b). The NC is distant from the channel by a tunnel oxide of thickness d_1 (chosen to be less than 4 nm in this study), from the gate by the control oxide

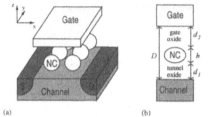

Figure 1. (a) Schematic view of a floating gate memory device. A NC layer is embedded in the insulator of the MOSFET. S and D stand respectively for source and drain electrodes. (b) Geometrical parameters of the single isolated NC used to model the whole layer.

of thickness d_2, and the NC height is noted h. In this work, the total thickness of the insulator D is fixed to 14 nm, but the position of the NC can vary, as well as h. To study the charging/discharging processes, the channel is grounded and a voltage V_g is applied to the gate.

In addition to geometrical fluctuations, the model allows to change easily the nature of materials by specifying the physical parameters (band edges, bandgap, effective mass, dielectric permittivity, ...). The NC is supposed to have an axial symmetry along a direction perpendicular to the plan of the reservoirs, but as long as this property is conserved, any NC shape can be considered. To keep this hypothesis valid, the physical parameters must have the same symmetry. Thus the effective mass is not forced to be isotropic, although initially the model was dedicated to NCs with unspecified crystalline orientation (in relation with NCs obtained by ion implantation).

THEORY

For this work we use an extension of our semi-analytical model which is already explained in detail in Ref. 10 and 11. The basis of the model is unmodified but the complete description of this new model will be found elsewhere [12].

In order to take into account the self-energy of carriers in the NC, we apply a rigid shift of the spectra computed in an empty NC. This shift is defined as the difference between the ground state of the empty NC and the self-consistent ground state of the NC with one electron [13]. In this approach, the wave functions used to compute the tunnel matrix elements are unchanged, nevertheless they have a physical meaning: Bardeen's formula can then be applied (which is not the case for the single particle wave functions obtained in the Kohn-Sham approach).

The major modification concerns the metallic electrode. One now considers arbitrary 2D subbands together with a specific density of states in order to recover the initial 3D density of the states associated with the metal electrodes. The gate electrode is then modeled in a similar way as the channel in the original model. Such a modification leads to a better understanding of the impact of the gate carriers on the structure. Indeed, for a given applied gate voltage V_g, we obtain the transition rates $W^{\pm}_{k\ np}(d_1, h)$ for every reservoir subband p, and every NC level n. The index k is set to 1 (resp. 2) if the motion of carrier is through the tunnel oxide (resp. control oxide), and $+$ (resp. $-$) stands for the motion orientation from channel to gate (resp. from gate to channel).

Since we use bulk values for our material parameters, we have introduced the effect of the image potential on the insulator band-edge, in order to simulate more realistic tunnel barriers. Thus we have adapted the resolution procedure of our finite element solver [14] to apply the method described in Ref. 15 and 16.

To compute the quantity of interest, we follow the same procedure as in our previous work [17]. The total transition rate of a given configuration $\{d_1, h\}$ is defined as

$$W_k^{\pm}(d_1, h) = \sum_{n,p} W_{k\,np}^{\pm}(d_1, h) \qquad (1)$$

Then the time for charging a single NC with one electron (from the channel) is given by $\tau_c = 1/W_1^+(d_1, h)$. In the case of a layer characterized by a set of N uncoupled NCs [18], $\{d_{1i}, h_i\}_{i=1..N}$, the charging time τ_c is deduced from the time needed for charging 95 % of the layer, i.e. $P(3\tau_c) = 0.95$, with

$$P(t) = 1 - \frac{1}{N} \sum_{i=1}^{N} \exp[-W_1^+(d_{1i}, h_i)\, t] \qquad (2)$$

By doing this for various sets of NC layers with a given size (number of NCs), we can extract statistical information about the charging time, while taking into account the geometrical fluctuations.

RESULTS AND DISCUSSIONS

In this work we use Si or Ge for the NC material, SiO_2 or HfO_2 as an insulator, a p-Si channel doped with $N_a = 10^{16}$ cm^{-3} acceptors, and a gold metallic gate. The table I summarizes the physical parameters used for the simulations. The geometrical parameters correspond to $D = 14$ nm, $h = 4$ nm, and d_1 in the range going from 1 to 4 nm (thus $d_2 = D - h - d_1$).

In figure 2 we present the band diagrams of various combinations of materials for $d_1 = 2$ nm. In the same diagram, we have plotted the electronic levels of the device. For the channel (resp. gate) green levels (resp. blue) are the energy bottoms of 2D subbands. In the NC, red levels are discrete energies in presence of one electron in the NC (dash red lines are the energies for the empty NC). Except for the hole consideration (not treated here), the only difference between the Si NC (fig. 2a) and the Ge NC (fig 2b) embedded in SiO_2 comes from the different electron effective mass. Thus as expected, only the number of levels in the NC changes (and their associated wave functions): the rest of the structure is unchanged. If the image potential is taken into account (fig. 2c), the levels of the Si NC are slightly lowered, as well as the relative height of the insulator barriers. By replacing the SiO_2 insulator by HfO_2, we observe that only few levels are conserved in the Si NC due to the low barrier height associated with HfO_2.

The influence of the band diagrams associated with different material combinations can be seen on the charging time τ_c . In figure 3, we plot the curves $\tau_c(V_g)$ computed for a range of gate bias V_g going from flat-band bias up to 5 V. The red line with crosses serves as a reference (Si NC in SiO_2). Adding the effect of image potential (green line with open square) shifts the peaks toward

Table I. Parameters of the various materials considered in this work.

Material	Rel. permittivity (ϵ_r)	Electr. affinity (χ)	Bandgap (E_g)	Eff. mass (m^*)
Si	11.7	4.01 eV	1.12 eV	0.27
Ge	16.2	4.00 eV	0.66 eV	1.64
SiO$_2$	3.8	1.10 eV	11.00 eV	0.50
HfO$_2$	20.0	3.30 eV	5.88 eV	0.15

31

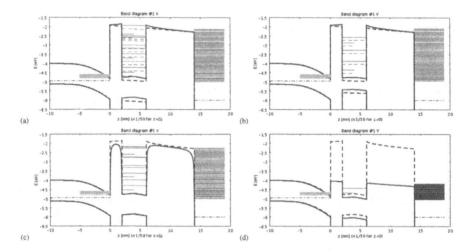

(a)   (b)   (c)   (d)

Figure 2. Electronic energy levels and band diagrams along the axis of axial symmetry ($d_1 = 2$ nm, $h = 4$ nm, and $d_2 = 8$ nm). Dash-dot lines stand for Fermi levels. (a) Si NC inside SiO_2 (dash lines are for empty NC). (b) Ge NC inside SiO_2 (dash lines are for empty NC). (c) Si NC inside SiO_2 with image potential (dash lines are for initial band-edges). (d) Si NC inside HfO_2 (dash lines are for SiO_2 case).

the low voltages (lowering of NC levels) and the charging time is diminished (lowering of the barrier height). By substituting Si with Ge in the NC (blue line with open triangle), the charging time is slightly increased as well as the interval between the peaks. It results only from the NC levels which are higher and more spaced for Ge NC: the 3rd level in Si and 2nd level in Ge are identical (fig. 2a-b) as seen in figure 3 between 2.5 V and 3.5 V. Finally, replacing SiO_2 by HfO_2 diminishes drastically the charging time since the barriers are 2.2 eV lower than those of SiO_2. Above 2.2 V, there is no more available level in the NC and the model is no more valid: indeed an electron tunnels directly to the gate.

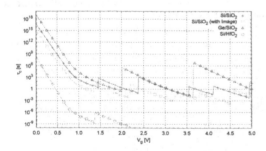

Figure 3. Charging time τ_c as a function of the gate bias V_g for the configurations of materials corresponding to the band diagrams shown in figure 2. The dimensions are fixed: $d_1 = 2$ nm, $h = 4$ nm, and $d_2 = 8$ nm.

32

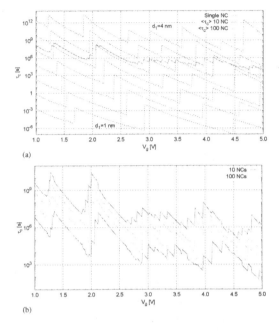

(a)

(b)

Figure 4. Charging time of Si NC inside SiO$_2$ *versus* V_g for fixed height h = 4 nm. For arrays of NCs, the position d_1 fluctuates according to a Gaussian distribution given by 2.5 ± 0.4 nm. (a) Blue dotted lines are for d_1 in the range [1;4] nm. The mean charging times $\langle \tau_c \rangle$ for 100 layers of 10 and 100 NCs are also shown (resp. in red and green). (b) Range of variation of the charging time around $\langle \tau_c \rangle$ as a function of the gate bias V_g, for 100 layers of 10 and 100 NCs. τ_c follows a log-normal law, thus the limits shown here correspond to $\langle \tau_c \rangle / \sigma_{\tau_c}$ and $\langle \tau_c \rangle \cdot \sigma_{\tau_c}$.

In what follows, only the oxide thickness d_1 is varied (the other geometrical parameters are fixed). We thus generate layers whose NCs are identical (same shape and same size h).

For a given material configuration, *i.e.* Si NC embedded in SiO$_2$, figure 4 provides some statistical information associated with a fixed height h of 4 nm for the NC. In figure 4a the curves of charging time corresponding to d_1 going from 1 nm to 4 nm are shown together with the mean charging time $\langle \tau_c \rangle$ of 100 layers composed of 10 or 100 random NCs (which are characterized by a Gaussian distribution $d_1 = 2.5 \pm 0.4$ nm). As expected, when d_1 is increased, the charging time of a single NC increases too. The peaks are shrunk and shifted toward lower voltages because of the change in electrostatic potential across the insulator. For layers of random NCs, the mean charging time is much more sensitive to the NCs with large d_1 values. Thus $\langle \tau_c \rangle$ is higher than the charging time associated with the mean value of d_1 (*i.e. 2.5 nm*). In figure 4b, the range of variation of the charging time for the 100 layers of 10 or 100 random NCs is shown. The limits plotted correspond to a standard deviation around $\langle \tau_c \rangle$. Since τ_c is found to follow a log-normal law, the limits are defined by $\langle \tau_c \rangle / \sigma_{\tau_c}$ and $\langle \tau_c \rangle \cdot \sigma_{\tau_c}$. The fluctuations of τ_c are sensitive to the size of the layer: larger is the layer, smaller are the variations.

CONCLUSIONS

In this paper, we have presented a model which allows to simulate a single NC inside an insulator and to obtain information about a layer composed of many single NCs. The geometrical parameters as well as the physical parameters can be adapted on request. Since the physical processes occurring between the channel, NCs, and gate have been explicitly highlighted, the charging of a NC layer can be controlled in a better way now. According to the statistical studies for charging times of NC layers, we are able to determine the expected fluctuations which are generated by geometrical fluctuations inherent to the elaboration process.

REFERENCES

[1] G. Molas, B. D. Salvo, D. Mariolle, G. Ghibaudo, A. Toffoli, N. Buffet, and S. Deleonibus, Sol. Stat. Elec. **47**, 1645–1649 (2003).

[2] S. Decossas, F. Mazen, T. Baron, G. Bremond, and A. Souifi, Nanotechnology **14**, 1272–1278 (2003).

[3] C. Bonafos, M. Carrada, N. Cherkashin, H. Coffin, D. Chassaing, G. B. Assayag, A. Claverie, T. Müller, K. H. Heinig, M. Perego, M. Fanciulli, P. Dimitrakis, and P. Normand, J. Appl. Phys. **95**(10), 5696–5702 (2004).

[4] M. Shalchian, J. Grisolia, G. B. Assayag, H. Coffin, S. M. Atarodi, and A. Claverie, Appl. Phys. Lett. **86**, 163111 (2005).

[5] C. M. Compagnoni, R. Gusmeroli, D. Ielmini, A. S. Spinelli, and A. L. Lacaita, J. Nanosci. Nanotechnol. **7**(1), 193–205 (2007).

[6] G. Iannaccone and P. Coli, Appl. Phys. Lett. **78**(14), 2046–2048 (2001).

[7] A. Thean and J. P. Leburton, IEEE Electron Dev. Lett. **22**(3), 148–150 (2001); J. S. de Sousa, A. V. Thean, J. P. Leburton, and V. N. Freire, J. Appl. Phys. **92**(10), 6182–6187 (2002).

[8] M. Prada and P. Harrison, New Journal of Physics **6**(1), 30 (2004).

[9] C. M. Compagnoni, D. Ielmini, A. S. Spinelli, and A. L. Lacaita, IEEE Trans. Elect. Dev. **52**(4), 569–576 (2005).

[10] B. Leriche, Y. Leroy, and A. S. Cordan, J. Appl. Phys. **100**(7), 074316–1/6 (2006).

[11] Y. Leroy, B. Leriche, and A. S. Cordan, Modeling transport in silicon nanocrystal structure, in *Proceedings of the COMSOL Multiphysics Conference 2005*, pp. 129–134, Paris, November 2005.

[12] Y. Leroy, D. Armeanu, and A. S. Cordan, (to be published).

[13] M. Dubois, S. Latil, L. Scifo, B. Grévin, and A. Rubio, J. Chem. Phys. **125**(3), 034708 (2006).

[14] Comsol AB, *Comsol Multiphysics Reference Manual, version 3.5*, 2008.

[15] J. G. Simmons, Image force in Metal-Oxide-Metal tunnel junctions, in *Tunneling phenomena in solids*, edited by E. Burstein and S. Lundqvist, chapter 10, pp. 135–148, Plenum Press, 1969.

[16] W. T. Norris, IEE Proc. Sci. Meas. Technol. **142**(2), 142–150 (1995).

[17] Y. Leroy and A. S. Cordan, Microelectron. Eng. **85**(12), 2354–2357 (2008).

[18] A. S. Cordan, Y. Leroy, and B. Leriche, Sol. Stat. Elec. **50**(2), 205–208 (2006).

Mater. Res. Soc. Symp. Proc. Vol. 1250 © 2010 Materials Research Society   1250-G01-05

Indium Gallium Arsenide Based Non-Volatile Memory Devices with Site-Specific Self-Assembled Germanium Quantum Dot Gate

Pik-Yiu Chan, Mukesh Gogna, Ernesto Suarez, Fuad Al-Amoody, Supriya Karmakar,
Barry I. Miller, John E. Ayers, and Faquir C. Jain

Department of Electrical and Computer Engineering, University of Connecticut, 371 Fairfield
Way, U-2157, Storrs, CT 06269, U.S.A.

ABSTRACT

This paper presents the implementation of indium gallium arsenide field-effect transistors
(InGaAs FETs) as non-volatile memory using lattice-matched II-VI gate insulator and quantum
dots of GeO_x-cladded Ge as the floating gate. Studies have been done to show the ability of II-
VI materials to act as a tunneling gate material for InGaAs based FETs, and GeO_x-cladded Ge
quantum dots having the ability to store charges in the floating gate of a memory device.
Proposed structure of the InGaAs device is presented.

INTRODUCTION

Recently, there have been many reports of implementing InGaAs FETs either on InP or
directly on Si substrates. Due to its high electron mobility, indium gallium arsenide high
electron mobility transistors (HEMTs) are one of the fastest transistors on the market. Unlike
HEMTs, InGaAs FETs in MOS or CMOS configuration require the use of a gate insulator that
provides minimum interface states resulting in reproducible threshold voltage behavior.
Finding an appropriate gate oxide for InGaAs based CMOS devices is challenging. Jain,
et. al. have shown that using high-k II-VI materials that are lattice-matched to the InGaAs
material serves as a good gate insulator for such devices [1].
Scaling of CMOS devices is also another problem currently faced by the memory
industry. Thinner gate material means faster operating speed of the memory device, but it can
also increase current leakage in the device. Memory devices with silicon nanocrystals as the
floating gate have long been introduced [2]. These silicon nanocrystals act as the charge storage
units, but retention of these charges become a problem. To solve this problem, cladded
nanocrystals have been used as the floating gate of the memory devices [3, 4].
This study combines the use of high-k II-VI materials as the gate insulator and the
cladded nanocrystals as the floating gate to create a new type of InGaAs non-volatile memory.

EXPERIMENTAL DETAILS

Figure 1 shows the cross-sectional schematic of the InGaAs non-volatile memory device.
The starting substrate was a 2000Å InGaAs layer grown on an InP substrate by metalorganic
chemical vapor deposition (MOCVD). The InGaAs layer was doped p-type by introducing zinc
during the growth. The deposition was done at 635°C, and the p-type doping was about
$10^{16}/cm^3$. 5000Å of PECVD oxide was deposited on the InGaAs layer to act as a masking oxide.
The pattern of the source and drain regions was formed, and the p-InGaAs regions were

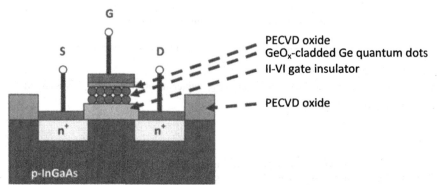

Figure 1. Cross-sectional schematic of the InGaAs non-volatile memory with GeO_x-cladded Ge quantum got as the floating gate.

etched away, and n-type InGaAs (doped with sulfur) was deposited by MOCVD at 635°C to form the n-type source and drain regions. The doping of the n-type source and drain regions was about $10^{18}/cm^3$.

The gate of this device is a stack of ZnS/ZnMgS/ZnS grown using photo-assisted MOCVD. A very thin layer of ZnSe, which is sandwiched between the p-InGaAs gate region and the II-VI gate insulator, acts as a buffer layer to enhance the adhesion of the gate stack. It was grown with metalorganic precursors dimethlyzinc (DMZn) and dimethlyselenium (DMSe) for 30 seconds at 505°C. For the gate insulator, ZnMgS, grown on top of the buffer layer, the precursors were DMZn at 43.6 Torr at 15 sccm; bismethyl-cyclopentadienylmagnesium [(MeCp)$_2$Mg] at 0.9 Torr at 10 sccm; and diethylsulfide (DES) at 60 Torr at 30 sccm. This growth was done at 333°C for 3 minutes at 35 mW/cm^2.

After the growth of the gate material, GeO_x-cladded Ge quantum dots were self-assembled over the gate region. Finely ball-milled Ge powder with 99.999% purity was kept in a pH-controlled, sealed oxidizing solution for two days under constant sonication. To obtain the final quantum dots, the solution underwent several centrifugations and etching process. Due to the nature of these quantum dots, only two layers of dots self-assembled over the p-type region of the InGaAs substrate, which is the channel region of the device, thus forming the floating gate of the memory device.

To make the device into a non-volatile memory, a thin layer (~75 Å) of PECVD oxide was deposited over the sample to act as the control gate of the device.

For the contacts of the devices, an alloy of gold, germanium and nickel (AuGeNi) was used as the contact metal for the source and the drain regions of the device, and aluminum was used as the gate metal contact.

DISCUSSION

A conventional InGaAs FET previously fabricated with lattice-matched II-VI gate insulator shows an output I_D-V_D characteristic as shown in Figure 2. The control of the lattice-matched gate on the device is shown by the increasing drain current with increasing gate voltage

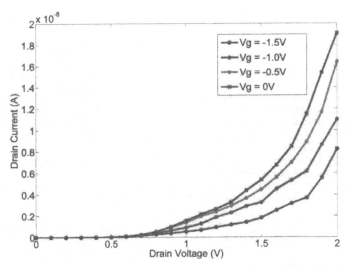

Figure 2. Output I_D-V_D characteristic of a conventional InGaAs FET with lattice-matched II-VI gate insulator.

at a given drain voltage. Gate insulator that is lattice-matched to the InGaAs material will have fewer interface states, which will reduce the fluctuation in the threshold voltage of the device.

The effect of GeO_x-cladded Ge quantum dots as the floating gate of a non-volatile memory device has been demonstrated [5]. Figure 3 shows a TEM image of a Ge nanocrystal with a 2nm diameter surrounded by a 1nm GeO_x cladding.

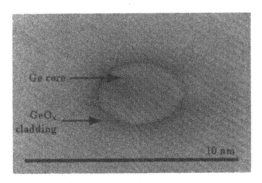

Figure 3. TEM image of a 2-nm Ge nanocrystal with 1-nm cladding.

A characteristic of a non-volatile memory is its ability to store charges. These charges will in turn change the threshold voltage of the device. For the proposed device with two layers

37

of GeO$_x$-cladded Ge quantum dots, the threshold shift, ΔV_{TH}, is given by equation 1 based on the model by Hasaneen, et. al. [6]:

$$\Delta V_{TH} = \frac{\int_0^{t_c} \left[q \cdot n_{ed} \cdot N_{qd} \cdot R_t \cdot A \right] dt}{C_{gd}} \tag{1}$$

where t_c is the charging time of the quantum dot gate during the write or program operation, q is the electron charge, n_{ed} is the number of electrons per dot, N_{qd} is the number of quantum dots in the floating gate, R_t is the tunneling rate probability of carriers from the channel to the quantum dots, A is the area of the active region, and C_{gd} is the capacitance between the gate and the quantum dots.

According to Hasaneen, et. al. [6], the tunneling rate probability, R_t, depends on the energies in the channel and in the quantum dots. The highest tunneling rate occurs when the quantum dot eigen energy is higher than the channel energy.

The charging mechanism of the memory device is due to hot electron injection. To inject charges, 6V is applied to the gate while the source is grounded. Short pulses of 10V are sent to the drain. These drain pulses cause the electrons to tunnel from the n-channel to either the top or the bottom layer of GeO$_x$-cladded Ge quantum dots. This mechanism is schematically represented in the energy band diagram in Figure 4.

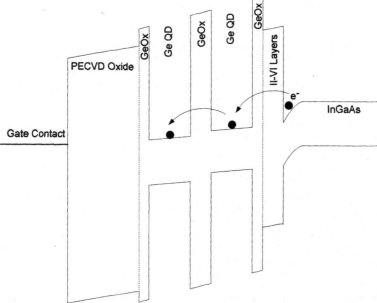

Figure 4. Schematic energy band diagram showing the charging mechanism of the InGaAs non-volatile memory.

To observe the shift in the threshold voltage due to charge injection, a silicon-based non-volatile memory with GeO_x-cladded Ge quantum dots was stressed by applying a 10-µs pulse of 10V to the drain, while maintaining a 6V at the gate and ground at the source. Figure 5 shows the transfer I_G-V_D characteristic before and after stress. The shift in the threshold voltage due to charge injection is about 0.3 V.

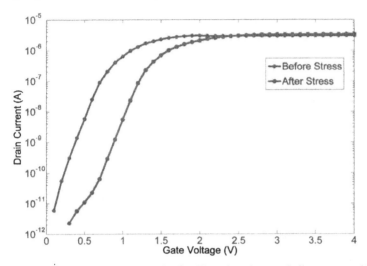

Figure 5. Transfer I_D-V_G characteristic of a silicon-based non-volatile memory with GeO_x-cladded Ge quantum dots as the floating gate, before and after stress.

The promising results of using lattice-matched II-VI material as the gate insulator in an InGaAs FET device and the charge storage property of GeO_x-cladded Ge quantum dots as the floating gate of a non-volatile memory device lead to the proposed structure of InGaAs based non-volatile memory with GeO_x-cladded Ge quantum dot as the floating gate. Results for this device are still pending at the time of submission of this paper.

CONCLUSIONS

The combination of using high-k II-VI materials that is lattice-matched to InGaAs as the gate insulator and utilizing GeO_x-cladded Ge quantum dots as the floating gate in an InGaAs-based non-volatile memory will combine the superior speed of the InGaAs devices and the high storage density of quantum dot gate memory. Realization of this device is in the process.

ACKNOWLEDGMENTS

The authors would like to express thanks for the support received from ONR contract N00014-06-1-0016 and NSF grant ECS-0622068, and discussions with Dr. D. Purdy (ONR).

REFERENCES

1. F.C. Jain, E. Suarez, M. Gogna, F. Al-Amoody, D. Butkiewicus, R. Hohner, T. Liaskas, S. Karmakar, P.-Y. Chan, B. Miller, J. Chandy, and E. Heller, *J. Electron. Mater.* **38**, 1574 (2009).
2. S. Tiwari, F. Rana, K. Chan, H, Hanafi, W. Chan, and D. Buchanan, *IEDM Proc.*, 521 (1995).
3. F.C. Jain, E. Heller, S. Karmakar, and J. Chandy, *Proc. International Semiconductor Device Research Symposium*, Dec. 12–15, College Park, MD (2007).
4. R. Velampati and F.C. Jain, *NSTI Nanotech*, Santa Clara, CA, May 20–24 (2007).
5. M. Gogna, F. Al-Amoody, S. Karmakar, F. Papadimitrakopoulos, and F. Jain, *Proc. Nanoelectron Devices for Defense and Security Conference*, Sept 28 – Oct 2, Fort Lauderdale, FL (2009)
6. E.-S. Hasaneen, E. Heller, R. Bansal, and F. Jain, *Solid State Electron.* **48**, 2055 (2004).

Mater. Res. Soc. Symp. Proc. Vol. 1250 © 2010 Materials Research Society   1250-G01-07

Ultra-Low Energy Ion Implantation of Si into HfO$_2$ and HfSiO-based Structures for Non Volatile Memory Applications

F. Gloux[1], P. E. Coulon[1], J. Groenen[1], S. Schamm-Chardon[1], G. Ben Assayag[1], B. Pecassou[1], A. Slaoui[2], B. Sahu[2], M. Carrada[2], S. Lhostis[3] and C. Bonafos[1]
[1] CEMES-CNRS and Université de Toulouse, nMat group, BP 94347, 31055 Toulouse cedex 4, France
[2] InESS – CNRS, 23 rue du Loess, 67037 Strasbourg, France
[3] ST Microelectronics, 850 rue Jean Monnet, 38926 Crolles, France

ABSTRACT

The fabrication of Si nanocrystals (NCs) in multilayer structures based on HfO$_2$ and alloys for memory applications is carried out using an innovative method, the ultra-low energy (1 keV) ion implantation followed by a post-implantation annealing. Si$^+$ ions are implanted into SiO$_2$ thin layers deposited on top of thin HfO$_2$-based layers. After annealing at high temperature (1050°C), the implantation leads to the formation of a two dimensional array of Si NCs at a distance from the surface larger than expected, due to an anomalous oxidation of the implanted Si. Nevertheless, the best memory windows are obtained at lower thermal budget, when no nanocrystals are present in the layer. This suggests that electrical measurements should always be correlated to structural characterization in order to understand where charge storage occurs.

INTRODUCTION

Nanocrystal memory (NCM) devices are competitive candidates for extending further the scalability of Flash-type memories [1-2]. Particularly, the fabrication of NCs into high-k dielectric matrices instead of SiO$_2$ has retained particular attention for achieving NCMs with low programming voltages and improved data retention. Among the different high-k materials under investigation, HfO$_2$ and its alloys are considered as very promising candidates for the integration in ultra-scaled commercial devices. Although promising device results using Si NCs embedded in HfO$_2$ gate dielectrics have been recently presented [3], the fabrication of semiconducting NCs in high-k materials does not remain straightforward because of diffusion and oxidation problems [4].

We have extensively used ultra-low energy ion implantation (ULE-II) to synthesize single planes of Si NCs embedded in very thin (5 to 10 nm) oxide layers. The depth-location of these two-dimensional (2D) arrays of particles below the surface can be controlled with nanometer precision by tuning the implantation energy while their size and density can be controlled by varying the dose and annealing conditions [5]. These parameters were finally optimized to fabricate non volatile memory devices with improved characteristics [2]. Recently, this approach was extended to synthesize a plane of Si NCs within thin HfO$_2$ and HfSiO layers deposited by metalorganic chemical vapour deposition (MOCVD) on Si wafers. A structural study by transmission electron microscopy (TEM) techniques revealed the complete oxidation of the implanted Si in the HfO$_2$-based layers while Si NCs were formed in the interfacial SiO$_2$ layer [6]. In this paper we investigate alternative structures consisting in a SiO$_2$ layer as Si NC host, deposited on top of a HfO$_2$-based layer used in this case as tunnel oxide only. Photoluminescence (PL) analysis is carried out to detect Si NCs via their emission in the

41

red/infra-red range. Structural and chemical analyses were carried out by TEM techniques at the nanometer scale, and electrical studies by capacitance-voltage (C-V) measurements.

EXPERIMENT

Nanocrystalline HfO_2 and HfSiO layers were grown by MOCVD on top of Si wafers. They are respectively 7 and 2 nm thick and are separated from the substrate by a thin 1 nm thick SiO_2 interfacial layer (IL). Either a 6 nm or 10 nm thick SiO_2 layer was deposited on top of HfO_2 and HfSiO layers by electron cyclotron resonance-chemical vapor deposition. Si^+ ions were implanted in the SiO_2 layers at a low energy of 1 keV and at a dose of $1.0 \times 10^{16} Si^+/cm^2$. The stacks have been further annealed at 1050 °C under N_2 ambient and for 60 s by rapid thermal annealing (RTA), for the purpose of Si-NCs formation. One of these samples has also been annealed at lower thermal budget (LTB) (950°C for 30s) still by RTA and another one at 1050°C but with conventional annealing (CA) during 30 minutes. Elaboration, implantation and annealing conditions are summarized in Table I.

Table I: Elaboration, implantation and annealing conditions of the investigated structures.

Deposited stack and name	Si$^+$ implantation: energy - dose	Annealing
SiO_2(10nm)/HfO_2(7nm)-LTB		950°C 30s
SiO_2(6nm)/HfO_2(7nm)		1050°C 60s
SiO_2(6nm)/HfSiO(2nm)	1.0keV-1.0×10^{16} at/cm^2	1050°C 60s
SiO_2(10nm)/HfO_2(7nm)		1050°C 60s
SiO_2(10nm)/HfSiO(2nm)		1050°C 60s
SiO_2(10nm)/HfSiO(2nm)-CA		1050°C 30min

Cross-section and plan view samples were prepared by mechanical polishing and ion milling. Conventional TEM was carried out with a Philips CM30 operating at 300kV while high resolution electron microscopy (HREM) and energy-filtered TEM (EFTEM) images were taken using a FEI Tecnai™ F20 operating at 200 kV, equipped with a spherical aberration corrector and a Gatan imaging filter. EFTEM was performed using the Si-plasmon energy at 17eV [7]. PL spectroscopy was carried out using a Dilor XY Raman spectrometer equipped with a super-notch filter eliminating the 488 nm excitation laser line, a 150 groves mm^{-1} grating and a liquid nitrogen-cooled CCD detector. The laser power at the entrance of the microscope was 150 µW or less. The incident and emitted lights were focused and collected by means of a confocal microscope equipped with a ×100 magnification objective working in air (0.95 numerical aperture). Capacitance-voltage (C-V) measurements were carried out on all the samples. Electrical characteristics were measured on 0.64 mm^2 MOS capacitors with Al gate electrodes as well as Al rear-side contact. C-V measurements were carried out using HP4192A impendence analyzer through a LABVIEW interface.

RESULTS AND DISCUSSION

All the samples annealed at high temperature (1050°C) present a typical PL band around 1.65-1.70 eV that does not exist in the Si bulk reference (see Figure 1). This feature is related to Si NCs formed in SiO_2 layers for similar implantation conditions and CA [2]. The PL band intensity is higher for the sample annealed in conventional furnace. In this case, healing of the non radiative defects is more efficient.

In order to confirm the nucleation of Si nanoparticles when annealing at high thermal budget, TEM observations have been carried out. A previous study proved that conventional TEM performed in Defocused Bright Field (DBF) conditions could successfully highlight the

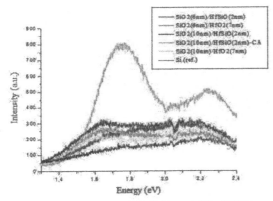

Figure 1: PL spectra of the SiO_2/HfO_2 and $SiO_2/HfSiO$ samples.

formation of a Si NC band in thin thermal SiO_2 layers [8] and allowed to measure characteristic distances with an error of ±0.5 nm. Figure 2.a represents a DBF image of the $SiO_2(10nm)/HfO_2(7nm)$ sample annealed at LTB (950°C for 30s). No Si NC band is visible. The LTB may be not sufficient for the phase separation and the Si NCs nucleation. Nevertheless, the final thickness of the SiO_2 layer (15 nm) after implantation and annealing is larger than the one expected after the incorporation of 10^{16} at/cm^2, which would represent 2 nm in terms of additional thickness. This can be due to a process of anomalous oxidation of the implanted Si that has already been observed [9] and will be discussed further. Figure 2.b is a DBF image of the $SiO_2(6 nm)/HfSiO(2 nm)$ sample annealed at 1050°C for 60s. In this case, we observe the presence of a band with a dark contrast inside the SiO_2 top layer, which is the projection of the 2D array of Si nanoparticles. All the samples annealed at 1050°C either by RTA or by CA exhibited a similar feature, showing that Si nanoparticles were formed. The crystalline nature of the nanoparticles was confirmed by HREM (not shown).

Statistical measurements of the characteristic distances and thicknesses were performed on DBF images on several areas for each sample. In particular, the position of the middle of the NC layer with respect to the surface (projected range), the thickness of the Si NC band, of the control oxide (CO) and of the SiO_2 layer, the distance between the NCs and the top of the high-k layer, and the total swelling of the SiO_2 layer were measured together with the thickness of the IL and of the high-k layer (Table II).

The projected range ranges roughly between 8 and 12.5 nm. It is far higher than the expected one at 4.2nm from similar ULE-II and CA conditions in SiO_2 matrices [5]. Moreover, the total thickness of the SiO_2 layer after implantation and annealing is about 15 nm for an initial layer of 6 nm and ranges between 15 and 20 nm for an initial layer of 10 nm. The total expansion is much larger than the expected one (2 nm) due to matter addition when implanting Si$^+$ with a dose of 10^{16}at/cm^2. An anomalous swelling of the SiO_2 layer ranging between 3 and 8 nm is thus observed. This unexpected expansion can be partly explained by the so-called "anomalous oxidation", already observed in Si implanted thin SiO_2 layers [9-12]. This "anomalous" oxidation takes place immediately after implantation, because the heavily damaged SiO_2 layers absorb humidity. The water molecules present in the air are driven in the layers, dissociate and finally react with the implanted Si to form SiO_2. During annealing, further dissociation of OH bonds takes place and finally most of the H atoms diffuse out to the surface. The penetration and final concentration of water molecules do not depend on the relative humidity in the atmosphere but are only limited by the degree of damage i. e., by the

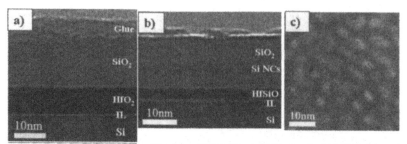

Figure 2: a) DBF image of the SiO_2(10nm)/HfO_2(7nm)-LTB sample; b) DBF image of the SiO_2(6nm)/HfSiO(2nm) sample; c) EFTEM image taken with the Si plasmon (17eV) on a plan view preparation of the SiO_2(6nm)/HfSiO(2nm) sample: Si NCs appear in white contrast.

concentration of defects, in the SiO_2 matrix. In the case of Si implanted in thermal SiO_2, we have evaluated that 40% of the implanted Si is oxidised. The remaining Si precipitates to form Si NCs. The presence of HfO_2-based layers in the stack, full of OH groups after deposition [13] can be another source of O_2. The swelling of the SiO_2 layer consecutive to the oxidation of the implanted Si takes place here in the control oxide (CO) because the Si implanted dose is very high. Indeed, in this case, a quasi-continuous layer of Si NC is formed, preventing the O_2 to go deeper and oxidise the Si substrate [2]. The thin IL between the Si substrate and the high-k measured after annealing (1 nm only) evidences that O_2 did not diffuse up to the substrate. For smaller doses, a swelling of the tunnel oxide (TO) is also expected [9]. Anyway, the whole oxidation of the implanted Si (2 nm in equivalent thickness) would lead to a swelling of about 5 nm, when considering the difference in volume between Si and SiO_2, and no nanocrystal would be present in the layer. This swelling is smaller than the ones measured here (see Table II). Thus the anomalous oxidation is not sufficient to explain the observed swelling and shift of the NCs position. The fact that we are dealing here with deposited SiO_2 which is in principle more porous than thermal silica, could be taken into account to explain the shift in position of the Si NCs compared to the calculated projected range of the implantation and with respect to the NCs position in thermal SiO_2 layers. Experiments are under progress in order to better understand the obtained morphology. Figure 2.c is an EFTEM image taken with the Si plasmon on a plan view preparation of the SiO_2(6nm)/HfSiO(2nm) sample, where the Si NC distribution in the plane can be directly investigated. In spite of the suspected Si NC oxidation a high NC density was measured, i.e.

Table II: Characteristic distances measured in DBF images for samples annealed at 1050°C.

Deposited stack and name	Projected range (nm)	Si NC band (nm)	CO (nm)	Distance NCs-high-k layer	Total SiO₂ layer (nm)	Total swelling (nm)	IL (nm)	High-k layer (nm)	Total equivalent thickness oxide (nm)
SiO_2(6nm) /HfO_2(7nm)	8.7	3.1	7.1	5.3	15.5	7.5	0.6	7.0	6.9
SiO_2(6nm) /HfSiO(2nm)	9.3	3.7	7.4	4.2	15.3	7.3	0.6	3.1	5.1
SiO_2(10nm) /HfO_2(7nm)	12.4	3.0	10.9	6.3	20.5	8.5	0.9	7.0	8.2
SiO_2(10nm) /HfSiO(2nm)	10.9	1.3	10.2	6.1	17.6	5.6	0.8	3.8	7.3
SiO_2(10nm) /HfSiO(2nm) -CA	7.9	1.9	5.9	7.8	15.6	3.6	1.3	3.1	9.4

$\sim 2 \times 10^{12}/cm^2$. The NC distribution exhibits a partial percolation supporting the assumption of an initial quasi-continuous layer of implanted Si, leading after oxidation to an increase of the CO thickness only. At the end, the thinnest distance between the NC band and the high-k layer top is obtained for the stack $SiO_2(6nm)/HfSiO(2nm)$ annealed at 1050°C for 60s, i. e. 4.2 nm. In this case the equivalent oxide thickness of the TO is EOT = 0.6 (IL) + 0.3 (HfSiO) + 4.2 (SiO_2) = 5.1 nm.

C-V measurements were carried out on all the samples. Only the LTB annealed sample shows the presence of a significant memory window. All other samples annealed at higher temperature (1050 °C) do not exhibit any hysteretic effect. Moreover, the C-V curves of these samples are not regular and show large frequency dispersion and stretch-out effect, indicating the presence of a large amount of interfacial traps. In order to get a clear picture, the C-V curves for the sample annealed at LTB and one sample annealed at high temperature (1050°C) are shown in Figures 3.a and b, respectively. No memory window is obtained on the samples containing the 2D array of Si NCs (Figure 3.b). As shown in a previous work [14], further annealing under oxidising ambient would be necessary in order to improve the quality of the CO and TO and obtain charge retention. On the contrary, the sample annealed at LTB, where no NCs are observed showed an interesting flat band voltage shift of 4 V for sweeping voltage of ±9 V (Figure 3.a). The direction of the observed hysteresis is counter-clockwise, indicating a net electron trapping in the MOS structure. The effect of mobile ions on charge trapping can be ruled out, since negligible counter-clockwise hysteresis in C-V curves was observed in the control sample without implantation or in the as-implanted sample. It is interesting to note that the frequency dependent C-V curves in the frequency range of 50 kHz to 1 MHz indicate negligible dispersion and/or stretch-out effect in the depletion region and the memory window remains almost constant. We speculate that the interface traps could not response to the applied ac signal in this frequency range. However, the increase of the hump in the lower part of the C-V curves with increasing frequency indicates the presence of considerable amount of slow traps in our stack structures. In this case, the charges are probably stored within defect related traps created during the implantation in the SiO_2 layer and not totally cured during the LTB annealing.

As a conclusion of this result, coupling C-V measurements to PL analysis and TEM observations is necessary for a better understanding of the charge storage properties.

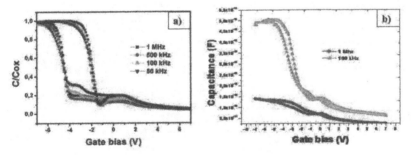

Figure 3: Typical frequency dependent C–V characteristics: a) for the sample annealed at LTB (950°C for 30 s); b) for the sample $SiO_2(6nm)$ /$HfO_2(7nm)$ annealed at 1050°C for 60s.

CONCLUSION

The structural and electrical properties of SiO_2 layers deposited on top of tunnel HfO_2-based layers, implanted with Si^+ at low energy then annealed, were analyzed in this work. When

45

annealing at high temperature (1050°C) either by RTA or by CA, a 2D array of Si NCs was successfully formed. An anomalous swelling was observed after implantation due to a partial oxidation of the implanted Si. This swelling takes place in the CO, leading to a shift in depth of the NC position compared to the expected values for the implantation projected range. Despite this anomalous oxidation, a high density of NCs is obtained in the plane, leading to partial percolation. C-V measurements show no significant memory window, despite the presence of a 2D array of Si NCs well located with respect to the Si substrate. On the contrary, interesting memory windows are obtained when annealing at LTB (950°C for 30s) where no NCs are formed. This study demonstrates the usefulness of coupling electrical measurements to rigorous TEM analysis for a better knowledge of the charge trapping.

ACKNOWLEDGEMENTS

This work is financed by the ANR project *"ANR/PNANO07-0053 – NOMAD"*.

REFERENCES

1. S. Tiwari, F. Rana, H. Hanafi, A. Hartstein, E. F. Crabbe and K. Chan, *Appl. Phys. Lett.* **68**, 1377 (1996).
2. C. Bonafos, H. Coffin, S. Schamm, N. Cherkashin, G. Ben Assayag, P. Dimitrakis, P. Normand, M. Carrada, V. Paillard and A. Claverie, *Solid-State Electronics* **49**, 1734 (2005).
3. J. Lu, Y. Kuo, J. Yan and C.-H. Lin, *Jap. J. Appl. Phys.* **45**, L901 (2006).
4. M. Fanciulli, M. Perego, C. Bonafos, A. Mouti, S. Schamm and G.Benassayag, *Adv. Sci. Technol.* **51**, 156 (2006).
5. C. Bonafos, M. Carrada, N. Cherkashin, H. Coffin, D. Chassaing, G. Ben Assayag, A. Claverie, T. Müller, K. H. Heinig, M. Perego, M. Fanciulli, P. Dimitrakis and P. Normand, *J. Appl. Phys.* **95**, 5696 (2004).
6. P. E. Coulon, K. Chan Shin Yu, S. Schamm, G. Ben Assayag, B. Pecassou, A. Slaoui, S. Bhabani, M. Carrada, S. Lhostis and C. Bonafos, *Materials Research Society Symposium Proceedings* **1160**, 3 (2009).
7. S. Schamm, C. Bonafos, H. Coffin, N. Cherkashin,1, M. Carrada,G. Ben Assayag, A. Claverie, M. Tencé and C. Colliex, *Ultramicroscopy* **108**, 346 (2008).
8. G. Ben Assayag, C. Bonafos, M. Carrada, A. Claverie, P. Normand and D. Tsoukalas, *Appl. Phys. Lett.* **82**, 200 (2003).
9. A. Claverie, C. Bonafos, G. Ben Assayag, S. Schamm, N. Cherkashin,V. Paillard, , P. Dimitrakis, E. Kapetanakis, D. Tsoukalas, T. Muller, B. Schmidt, K. H. Heinig, M. Perego, M. Fanciulli, D. Mathiot, M. Carrada and P. Normand, *Diffusion in Solids and Liquids* **258-260**, 531 (2006).
10. J. P. Biersack and L. G. Haggmark, *Nucl. Instrum. Methods* **174**, 257 (1980)
11. M. Carrada, N. Cherkashin, C. Bonafos, G. Benassayag, D. Chassaing, P. Normand, D. Tsoukalas, V. Soncini and A. Claverie, *Mat. Sci. And Eng.* **B101**, 204 (2003).
12. B. Schmidt, D. Grambole and F. Herrmann, *Nucl. Instr. and Meth. in Phys. Res.* **B191**, 482 (2002).
13. C. Wiemer, L. Lamagna, S. Baldovino, M. Perego, S. Schamm-Chardon, P. E. Coulon, O. Salicio, G. Congedo, S. Spiga and M. Fanciulli, *Appl. Phys. Lett.* **96**, 182901 (2010).
14. P. Normand, E. Kapetanakis, P. Dimitrakis , D. Tsoukalas, K. Beltsios, N. Cherkashin, C. Bonafos, H. Coffin, G. Benassayag and A. Claverie, *Appl. Phys. Lett.* **83**, 168 (2003).

Charge Trapping Sites in nc-RuO embedded ZrHfO High-k Nonvolatile Memories

Chen-Han Lin and Yue Kuo

Thin Film Nano and Microelectronics Research Laboratory, Texas A&M University, College Station, TX 77843, USA

ABSTRACT

Materials and electrical properties of the MOS capacitor containing nc-RuO embedded in the high-k dielectric ZrHfO film have been studied. The electron- and hole-trapping capacities and trapping sites in this kind of device were investigated using the constant voltage stress method, the frequency-dependent C-V measurement, and the retention characteristics. The negligible charge trapping phenomenon in the non-embedded device rules out the possibility of any trapping site in the bulk ZrHfO film or at the Si/ZrHfO interface. The electrical characterization results suggest that electrons are trapped in the bulk nc-RuO. However, holes have two possible trapping sites, i.e., in the bulk nc-RuO or at the nc-RuO/ZrHfO interface.

INTRODUCTION

The MOS capacitor containing the nanocrytstals embedded high-k dielectric is a promising flash memory device that can be scaled down to a very small geometry. For the conventional polycrystalline Si (poly-Si) floating gate structure, the stored charges could be easily drained off via a single leakage path formed in the tunnel oxide. This is an important reliability issue, which can be improved by replacing the continuous poly-Si film with nanocrystals [1]. Nanocrystals can be made of various materials, e.g., semiconducting elements (Si, Ge [1-3]) and oxides (ZnO [4]), conducting metals (Au, Pt, Ag, W, etc. [5-8]) and oxides (Sn-doped indium oxide, RuO_2 [9,10]), and insulating compounds (HfO_2 [11]). The memory functions and reliabilities of these devices have been studied. Nanocrystals with large work functions are ideal materials for the memory application because they can serve as the deep charge trapping sites to achieve a long data retention time. Ruthenium oxide (RuO) has a large work function of ~5eV and excellent thermal and chemical stability [12], which is suitable for the above application.

High-k dielectrics can replace the thermally grown SiO_2 as the tunnel and control oxides in the nanocrystals embedded memory devices [13]. Based on the same equivalent oxide thickness (EOT), the high-k film is physically thicker than SiO_2, which favors the charge retention time and is more reliable. HfO_2 is a popular high-k for gate dielectric application because it has a large band gap (~6eV) and low conduction and valence band offsets with Si [14]. The nanocrystals embedded HfO_2 memory device has been fabricated [15]. However, a high temperature annealing, e.g., >700°C is required to form nanocrystal and to eliminate defects in the HfO_2 film. It causes crystallization of HfO_2, which is a reliability concern. Previously, the authors proved that the amorphous-to-polycrystalline temperature of the sputter deposited HfO_2 could be increased by adding Zr into the film. In addition, compared with the un-doped HfO_2 film, the Zr-doped HfO_2 film (ZrHfO) can be prepared into a capacitor with a lower EOT, a thinner interface thickness, and a lower interface state density [16]. Therefore, the nanocrystalline RuO (nc-RuO) embedded ZrHfO is useful for high-capacity flash memories. Our previous study showed that both electrons and holes could be trapped to this kind of device depending on the polarity and magnitude of the gate bias [10]. Maikap et al. also reported a nc-RuO$_x$ embedded high-k memory composed of the Si/SiO$_2$/HfO$_2$/nc-RuO$_x$/Al$_2$O$_3$ structure [17]. Although both studies demonstrated the charge trapping phenomena, there was no detailed study on where charges were trapped. It is imperative to identify the exact sites responsible for charge traps. In this paper, authors investigated the electron- and hole-trapping characteristics of the MOS capacitor that contains the nc-RuO embedded ZrHfO high-k dielectric film.

EXPERIMENTAL

The dilute HF cleaned p-Si (100) wafer (dopant concentration 10^{15} cm^{-3}) was used as the substrate. The bottom ZrHfO (tunnel oxide)/Ru/top ZrHfO (control oxide) thin film stack was sequentially deposited by RF magnetron sputtering at 5 mTorr in one pump down process without breaking the vacuum. The bottom and top ZrHfO films were deposited using a Zr/Hf (wt% 12/88) composite target at 100W in an Ar/O$_2$ (1:1) ambient for 2 min and 4 min, respectively. The Ru film was prepared from a Ru target at 80W in the Ar ambient for 1 min. The expected thickness of as-deposited ZrHfO tunnel oxide, Ru layer, and ZrHfO control oxide should be 2nm, 3nm, and 4nm, respectively. The post deposition annealing (PDA) step was done at 950°C for 1 min in a N$_2$/O$_2$ (1:1) ambient using a rapid thermal annealing (RTA) equipment (Modular Process Technology RTP-600S). This annealing process transformed the as-deposited Ru film into discrete nc-RuO with a 2-D spatial density of 8x10^{11} cm^{-2} [10]. The Ru-to-RuO transformation is confirmed from the selected area electron diffraction (SAD) patterns, as shown in Figures 1 (a) and (b).

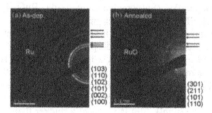

Fig. 1. SAD patterns of (a) as-deposited Ru and (b) after-annealed nc-RuO samples.

For the MOS capacitor fabrication, a 120 nm thick Al film was sputter-deposited on top of the high-k stack and subsequently defined and etched into gate electrodes (7.85x10^{-5} cm^2). The Al film was also deposited on the backside of the wafer for ohmic contact. The complete capacitor was annealed at 300°C for 5 min under the H$_2$/N$_2$ (10/90) atmosphere. The control sample, i.e., without the embedded nc-RuO, was prepared under the same condition. The capacitor's capacitance-voltage (C-V) curve was measured using an Agilent 4284A LCR meter. The flatband voltage (V$_{FB}$) was extracted from the C-V curve using the NCSU CVC program [18].

RESULTS AND DISCUSSION

Figure 2 shows the cross-sectional TEM micrograph (JEOL 2010 at 200keV accelerating voltage) of the nc-RuO embedded ZrHfO high-k stack after PDA. It clearly shows 5-7nm nc-RuO dots dispersed in the ZrHfO matrix. In a previous study, authors had observed from the XRD pattern measurement that after PDA, the pure ZrHfO film remained amorphous [19]. The lattice fringe space of 0.317nm in the nanocrystal of Fig. 2 is consistent with that of the RuO (110) structure (JCPDS 40-1290). The thickness of the ZrHfO tunnel oxide is 1.2 nm ~1.58 nm. The thickness of the RuO nc layer is 3.82 nm ~5.21 nm. The thickness of the ZrHfO control oxide is 1.39 nm ~2.12nm. In addition, a 1.7 nm AlO_x interface layer was formed between the Al gate and the top ZrHfO layer, which has a high barrier height of 2.9 eV. Therefore, the total control oxide thickness is > 3 nm with a high charge barrier height to the Al gate. A 4.8nm interface layer is formed between the ZrHfO film and the Si wafer. The energy dispersive spectrum (EDS) data, as show in the inset of Fig. 2, confirm our early observation that this interface layer is a Hf-silicate [20]. Although this interface layer may decrease the electric field across the ZrHfO/nc-RuO/ZrHfO stack and result in the less effective electronic coupling between nc-RuO and the Si substrate, this high-k film has lower electron and hole energy barriers than those of SiO_2 [21], which benefits the data retention and programming efficiency

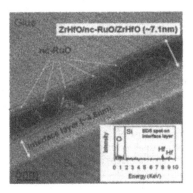

Fig. 2. Cross-sectional TEM view of nc-RuO embedded ZrHfO stack. The inset is the EDS pattern of the interface layer.

In order to distinguish the electron- and hole-trapping characteristics in the nc-RuO embedded ZrHfO MOS capacitor, the constant voltage stress (CVS) study was carried out by applying a gate voltage (V_g) of +9V and -9V for 10 s, separately. Before the gate voltage stress was applied, the capacitor's C-V curve was measured at 1MHz over a small V_g range of -2.5V to 0.5V. This small V_g sweep introduces negligible amount of charges to the dielectric layer. Therefore, the C-V shift after the CVS is due to

49

charge trapping. Figure 3(a) shows that the control sample has the very limited charge storage capability after +9V or -9V stress. Therefore, neither the bulk ZrHfO nor the Si/ZrHfO interface serves as the charge trapping site. However, Figure 3(b) shows that the nc-RuO embedded ZrHfO capacitor has V_{FB} shift (ΔV_{FB}) of -1V and 0.58V after -9V and +9V CVS, respectively. This corresponds the hole- and electron-trapping densities (Q's) of 2.24×10^{12} cm^{-2} and 1.29×10^{12} cm^{-2}, respectively, according to the equation of

$$Q = \frac{C_{acc}\Delta V_{FB}}{q}$$

where C_{acc} is the device's unit capacitance at the accumulation regime ($\sim 3.82\times10^{-7}$ F/cm^2) and q is the electron charge. The nc-RuO embedded device traps more holes than electrons under the same magnitude of stress voltage and time. There are several possible explanations, such as different electron- and hole-trapping sites, charge supply concentrations at the Si surface, or deep traps aligned in the nc-RuO bandgap.

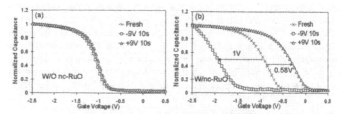

Fig. 3. C-V curves (a) control sample and (b) nc-RuO embedded ZrHfO capacitor before (fresh) and after CVS at ±9V for 10s.

Since the control sample has poor charge trapping capability, as shown in Fig. 3(a), the bulk nc-RuO or the ZrHfO/nc-RuO interface are possible charge-trapping sites in the nc-RuO embedded ZrHfO capacitor. In order to differentiate the hole- and electron-trapping sites, the frequency-dependent (from 100kHz to 1MHz) C-V measurement was carried out on the nc-RuO embedded ZrHfO capacitor and the control sample. Figure 4 shows two groups of C-V curves, i.e., from -9V to +9V for the hole trapping study and from +9V to -9V for the electron trapping study. Only the trapped holes, not the trapped electrons, can respond to the low measuring frequency. The insets in Fig. 4 show that the control sample has no frequency-dependency on electron- or hole-trapping C-V curves. Therefore, the interface states located at the Si/Hf-silicate/ZrHfO site are not responsible for the frequency-dependent C-V shift. For the nc-RuO embedded capacitor, the hole-trapping C-V curve was stretched and the V_{FB} shifted positively when the measure frequency was lowered. It was reported that charges stored at the nanocrystals/surrounding oxide interface may respond to the low measurement frequency [22]. In the nc-RuO embedded device, holes trapped at the nc-RuO/ZrHfO interface may tunnel toward the Si substrate when the band structure turns to the flat band condition. This is consistent with the observation that the hole-trapped C-V curve shifts to the positive V_g direction and a hump is formed near the V_{FB} when the measurement frequency is low. Therefore, in the nc-RuO embedded capacitor, holes may be trapped at two sites, i.e., in the bulk nc-RuO or at the nc-RuO/ZrHfO interface. This might also explain why more holes than electrons can be stored in the device under the same magnitude of gate stress voltage. In contrast to the hole-trapped C-V curves, there is no frequency-dependency on the electron-trapped C-V curve. This implies that electrons are deeply trapped in the bulk nc-RuO site.

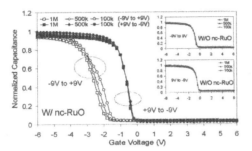

Fig. 4. Frequency-dependent C-V curves of nc-RuO embedded ZrHfO capacitor and control sample (inset). V_g was swept from –9V to +9V and from +9V to –9V for hole- and electron-trapped characterization, respectively.

Figure 5 shows the loss of trapped electrons and holes in the nc-RuO embedded ZrHfO capacitor as a function of retention time. Electrons and holes were injected into the device by the same CVS conditions as those in Fig. 3, i.e., +9V and -9V for 10s, respectively. Majority of the trapped electrons remain in the device for a long period of time. However, a portion of the trapped holes (~20% of total amount) were quickly lost within the first 50s. The quick loss of trapped holes in the nanocrystals-embedded memory device was also reported in another study [23]. The difference in electron and hole retention characteristics is consistent with the previous conclusion that electrons are deeply trapped in the bulk nc-RuO site but holes can be either deeply trapped in the nc-RuO site or loosely trapped to the ZrHfO/nc-RuO interface.

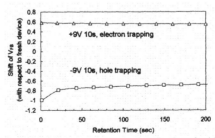

Fig. 5. Electron and hole retention characteristics of nc-RuO embedded ZrHfO capacitor.

CONCLUSIONS

The nc-RuO embedded ZrHfO capacitor has been fabricated and studied for its charge trapping characteristics. The formation of nc-RuO dots in the ZrHfO high-k film has been confirmed by TEM. The MOS capacitor containing only the ZrHfO dielectric film has negligible charge trapping capability. However, both electrons and holes can be stored in the nc-RuO embedded ZrHfO capacitor. There is no charge trapping site in the bulk ZrHfO film or at the Si/ZrHfO interface. The frequency-dependent C-V measurement show that the trapped holes, not the trapped electrons, respond to the low measurement frequency. There are different trapping sites for electrons and holes. Holes and electrons are trapped to the bulk nc-RuO site, which are insensitive to the measurement frequency. Holes can also be trapped to the ZrHfO/nc-RuO interface, which are easily tunnel back to the Si substrate.

ACKNOWLEDGMENTS

This project is partially supported by NSF CMMI 0926379 project. Chen-Han Lin thanks Applied Materials for providing AMAT PhD fellowship.

REFERENCES

1. S. Tiwari, F. Rana, H. Hanafi, A. Hartstein, E. F. Crabbé, and K. Chan, in *Appl. Phys. Lett.*, vol. 68, p. 1377, 1996.
2. H. I. Hanafi, S. Tiwari, and I. Khan, in *IEEE Trans. on Electron Devices*, vol. 43, p. 1553, 1996.
3. M. Kanoun, A. Souifi, T. Baron, and F. Mazen, in *Appl. Phys. Lett.*, vol. 84, p. 5079, 2004.
4. C.-H. Lin and Y. Kuo, in *Electrochem. Solid-State Lett.*, vol. 13, p. H83, 2010.
5. C.-C. Wang, J.-Y. Wu, Y.-K. Chiou, C.-H. Chang, and T.-B. Wu, in *Appl. Phys. Lett.*, vol. 91, p. 202110, 2007.
6. J.-Y. Tseng, C.-W. Cheng, S.-Y. Wang, T.-B. Wu, K.-Y. Hsieh, and R. Liu, in *Appl. Phys. Lett.*, vol. 85, p. 2595, 2004.
7. S.-W. Ryu, C. B. Mo, S. Y. Hong, and Y.-K. Choi, in *IEEE Trans. on Nanotechnology*, vol. 7, p. 145, 2008.
8. S. K. Samanta, W. J. Yoo, G. Samudra, E. S. Tok, L. K. Bera, and N. Balasubramanian, in *Appl. Phys. Lett.*, vol. 87, p. 113110, 2005.
9. A. Birge, C.-H. Lin, and Y Kuo, in *J. Electrochem. Soc.*, vol. 154, p. H887, 2007.
10. C.-H. Lin and Y. Kuo, in *ECS Trans.*, vol. 13, p. 465, 2008.
11. Y.-H. Lin, C.-H. Chien, C.-T. Lin, C.-Y. Chang, T.-F. Lei, in *IEEE Trans. on Electron Devices*, vol. 53, p. 782, 2006.
12. L. Krusin-Elbaum and M. Wittmer, in *J. Electrochem. Soc.*, vol. 135, p. 2610, 1988.
13. J. J. Lee, X. Wang, W. Bai, N. Lu, and D. L. Kwong, in *IEEE Trans. on Electron Devices*, vol. 50, p. 2067, 2003.
14. G. D. Wilk, R. M. Wallace, and J. M. Anthony, in *Appl. Phys.*, vol. 89, p. 5243, 2001.

15. L. Kang, B. H. Lee, W.-J. Qi, Y. Jeon, R. Nieh, S. Gopalan, K. Onishi, and J. C. Lee, in *IEEE Electron Device Lett.*, vol. 21, p. 181, 2000.
16. J. Lu, J, Yan, S. Chatterjee, H. C. Kim, and Y. Kuo, in 6th AVS Symposium and *International Conference on Microelectronics and Interfaces*, p. 47, 2005.
17. S. Maikap, T. Y. Wang, P. J. Tzeng, C. H. Lin, L. S. Lee, J. R. Yang, and M. J. Tsai, in *Appl. Phys. Lett.*, vol. 90, p. 253108, 2007.
18. J. Hauser and K. Ahmed, in *Characterization and Metrology for ULSI Technology*, p. 235, AIP, New York, 1998.
19. C.-H. Lin and Y. Kuo, in *ECS Trans.*, vol. 16, p. 309, 2008.
20. J. Yan, Y. Kuo, and J. Lu, in *Electrochem. Solid-State Lett.*, vol. 10, p. H199, 2007.
21. H. Jin, S. K. Oh, H. J. Kang, and M.-H. Cho, in *Appl. Phys. Lett.*, vol. 89, p. 122901, 2006.
22. S. Huang, S. Banerjee, R. T. Tung, and S. Oda, in *J. Appl. Phys.*, vol. 93, p. 576, 2003.
23. J.-S. Lee, Y.-M. Kim, J.-H. Kwon, H. Shin, B.-Y. Sohn, and J. Lee, in *Adv. Mater.*, vol. 21, p. 178, 2009.

Mater. Res. Soc. Symp. Proc. Vol. 1250 © 2010 Materials Research Society
1250-G01-09

Co/HfO$_2$ Core-Shell Nanocrystal Memory

Huimei Zhou[1], James A. Dorman[2], Ya-Chuan (Sandy) Perng[2], Stephanie Gachot[2], Jian Huang[1],
Yuanbing Mao[2], Jane P. Chang[2] and Jianlin Liu[1]
[1]Department of Electrical Engineering, University of California, Riverside, California 92521
[2]Department of Chemical Engineering, University of California, Los Angeles, California 90095

ABSTRACT

Metal/high-k dielectric core-shell nanocrystal memory capacitors were demonstrated.
This kind of MOS memory shows good performance in charge storage capacity, programming
and erasing speed. By using a self-assembled di-block co-polymer, Co/HfO$_2$ core-shell
nanocrystals showed uniform size and inter distance between crystals. Compared with traditional
metal nanocrystal fabrication process with E-Beam Evaporation followed by RTA (Rapid
Thermal Annealing), core-shell nanocrystal memory prepared by the co-polymer process
produces a wide memory window of 8.4V at the ±12 V voltage sweep. Co/HfO$_2$ core-shell
nanocrystals prepared by the low-temperature co-polymer process ensure high reliability of the
devices.

INTRODUCTION

Nanocrystal (NC) floating gate memory devices have received considerable attention due
to its excellent memory performance and high scalability [1-4]. In this kind of memory structure,
discrete trapping is used to store charges, which improve charge loss ratio encountered in
conventional flash memories. Several approaches, such as deeper well nanocrystals [5,6],
dielectric nanocrystals[7,8], and double-layer nanocrystals [9,10] were tried to obtain wider
memory window, longer retention time, and faster writing/erasing speed. Metal nanocrystals
have larger work function than that of silicon nanocrystals, and are advantageous to reduce the
leakage current through the tunneling barrier due to the increased barrier height [12]. These
metal particles were demonstrated to achieve better memory performance [10-12]. To avoid
reaction between the metal nanocrystal and oxide layer, silicide coated hetero-structure
nanocrystals were also proposed, with the goal of prolonging the retention time [13]. At the same
time, different high-k materials such as HfO$_2$, Al$_2$O$_3$ were used as the tunneling oxide to improve
the retention performance of the memory devices [14-16]. In this paper, we report our novel
structure of using core-shell nanocrystals for charge storage, which will improve the retention
time, and programming/erasing (P/E) performances. We also demonstrate metal/high-k core-
shell nanocrystal memory through a low temperature co-polymer process.

EXPERIMENT

For comparison, the nanocrystals were prepared in two ways. First, a 5.0-nm-thick
thermal oxide was grown in dry oxygen at 850^{0}C. A thin HfO$_2$ layer of around 3nm was then
deposited on SiO$_2$ layer by ALD (Atomic Layer Deposition). An ultra-thin (~2nm) blanket Co
layer was deposited through e-beam evaporation followed by RTA (Rapid Thermal Annealing)

in N_2 at 650^0C to form nanocrystals. Another thin HfO_2 layer of around 3nm was deposited again on the nanocrystals to create Co/HfO_2 core-shell structure. Control oxide of about 15 nm was then deposited by LPCVD (Low Pressure Chemical Vapor Deposition). After control gate pattern formation, the control gate and backside gate were formed to make a MOS structure memory capacitor.

A self assembly di-block co-polymer process was used to deposit a highly ordered nanocrystal array. First, a 5.0-nm-thick thermal oxide was grown in dry oxygen. A thin HfO_2 layer of around 3nm was then deposited on SiO_2 layer by ALD. Self-assembly of cobalt was achieved by mixing PS-b-P4VP, toluene and $CoCl_2 \bullet 6H_2O$ followed by spin-coating onto the sample surface. The polymer was subsequently removed by exposure to an oxygen plasma for 3 min. The ordered Co particles were formed after a forming gas annealing at 400^oC for 30 minutes to reduce the surface cobalt oxide to cobalt [20]. A thin HfO_2 layer of ~3nm was deposited again on Co nanocrystals to form core-shell structures. Control oxide of about 15 nm was then deposited by LPCVD and MOS structure memory capacitor was fabricated.

RESULTS AND DISCUSSION

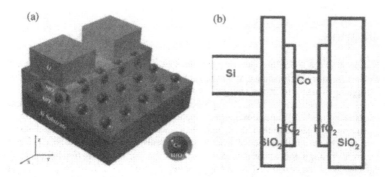

Fig. 1 a) Device structure of Co/HfO_2 core-shell dot memory, b) band diagram of Co/HfO_2 core-shell dot memory.

Figures 1 (a) and (b) show the 3-D (3 Dimensional) schematic and energy band diagram of Co/HfO_2 core-shell nanocrystal memory, respectively. In Fig. 1 (a), the core-shell nanocrystals are well ordered. The core-shell structure is represented by orange spheres with a transparent red shell. The band alignment of the core-shell nanocrystal memory along z direction is indicated in Fig. 1 (b) [17-18]. Similar to the hetero-structure nanocrystals, HfO_2 acts as a barrier, increasing the charge trapping ability and resulting in longer retention times [19].

Figure 2 (a) shows the 2D diagram of core-shell structure. A single electron is located in the center nanocrystal while the neighboring structures are empty. The potential distribution around the electron is calculated semi-classically. According to the following equations $\oint_s \varepsilon \vec{E} \bullet d\vec{S} = Q$ and $\phi = -\int_{l_0}^R \vec{E} d\vec{L}$, the potential distribution along the red line in Fig. 2 (a) is shown in Fig. 2 (b). The red line in Fig. 2 (b) shows the potential distribution of the core-shell structure. Potential distribution of the nanocrystal without shell is represented by the blue curve in (b). It is clearly seen that in the nanocrystal region, the potential of the core-shell structure is lower than the nanocrystal without shell. This means the shell structure screens the potential and reduces the Coulomb Blockade effect between particles. As a result of screening, the core-shell structure increases the electrons stored per nanocrystal.

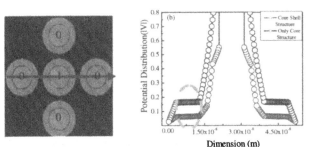

Fig.2 a) 2D diagram of core-shell structure, an electron is set in the center nanocrystal while surrounding nanocrystals remain empty, b) Potential distribution of the core-shell structure along red line of a), Potential difference highlighted in green shows that the core-shell structure reduces the Coulomb Blockade effect between particles.

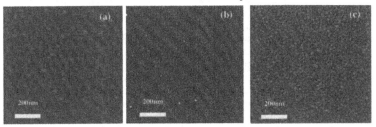

Fig. 3 AFM images for a) Co/HfO$_2$ core-shell particles by e-beam evaporation, b) Co co-polymer particles, and c) Co/HfO$_2$ core-shell particles by co-polymer.

Figures 3 (a), (b) and (c) show the AFM (Atomic Force Microscope) images of the Co/HfO$_2$ core-shell nanocrystals prepared by e-beam evaporation, Co nanocrystals and Co/HfO$_2$ core-shell nanocrystals prepared by the co-polymer process respectively. Fig. 3 (a) shows a non-uniform nanocrystal size ranging between 3 - 12nm. Fig. 3 (b) shows uniform nanocrystal size and spacing between particles. The size of nanocrytals is ~10nm with a spacing of ~8nm. After the HfO$_2$ layer deposition, the AFM images still show uniform particle size and spacing, which is indicated in Fig. 3 (c).

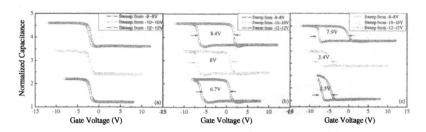

Fig. 4 C-V sweep memory window for a) capacitor with only HfO_2 shell, b) capacitor with Co/HfO_2 core-shell particles by di-block co-polymer, and c) capacitor with Co/HfO_2 core-shell particles by e-beam evaporation.

The capacitors were characterized by Agilent LCR meter at room temperature. Figures 4 (a), (b) and (c) show typical high frequency (1MHz) capacitance-voltage (C-V) sweep results with a scanning range between ±8V to ±12V for the memory capacitors with only HfO_2 shell, Co/HfO_2 core-shell structure prepared by co-polymer, and Co/HfO_2 core-shell structure prepared by e-beam evaporation, respectively. The sweep was run from the inversion to accumulation region and back with a sweep rate of 0.5V/s.

In Figure 4 (a), when voltage is swept from -8~8V, -10~10V and -12~12V, a tiny memory window (~0.01V at ±12V sweep) is shown which may be caused by defects in the HfO_2 layer. Fig. 4 (b) exhibits an obvious hysteresis as the gate voltage is increased. A memory window of ~6.7V is observed as the voltage is swept at ±8 V. When the sweep voltage is increased to 10V and 12 V, the memory window is increased at 8V and 8.4V, respectively. Wider voltage sweep range leads to the fact that more electrons are programmed to the nanocrystals and erased from the nanocrystals, therefore larger memory window is achieved. A similar response is observed in the Co/HfO_2 core-shell structure prepared by e-beam evaporation. In Fig.4 (c), when voltage is swept from -8~8V, -10~10V and -12~12V, it shows an increasing memory window of 2.5V, 3.4V and 7.9V, respectively. This means more and more electrons are programmed and erased with the increase of sweeping gate voltage.

Figure 5 (a) shows the flat band voltage shift (ΔV_{FB}) as a function of programming and erasing (P/E) time in co-polymer based Co/HfO_2 core-shell nanocrystal memory capacitor. It is evident that ΔV_{FB} increases with P/E time until saturation occurs. This is due to the fact that as the programming time increases, more and more electrons are injected into the nanocrystals until they are unable to accept more electrons. The same situation is observed as the device is erased. When increasing the erasing time, more and more electrons are erased until saturation. It is found that when the P/E voltage is increased, more electrons can go through the tunneling layer by F-N tunneling and are injected to nanocrystals or erased from nanocrystals. The higher P/E voltage is, the larger flat band voltage shift is achieved. A similar case happened in the core-shell nanocrystals prepared by e-beam evaporation, as shown in Fig. 5 (b). The magnitude of ΔV_{FB} increases as the P/E time and voltage increase until saturation.

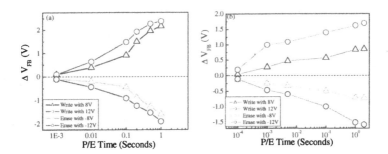

Fig. 5 Writing/Erasing performance for capacitors with Co/HfO$_2$ core-shell particles, (a) by co-polymer, and (b) by e-beam evaporation.

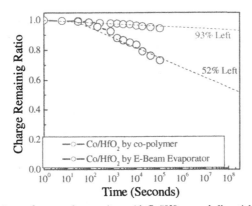

Fig. 6 Retention performance for capacitors with Co/HfO$_2$ core-shell particles by co-polymer and e-beam evaporation

The retention characterization is shown in Fig. 6 for the two capacitors with Co/HfO$_2$ core-shell nanocrystal memory capacitors. The capacitors were programmed at 15 V for 1 second. The co-polymer based Co/HfO$_2$ core-shell nanocrystal memory capacitor leads to a slower charge loss ratio. The evaporated Co/HfO$_2$ based capacitor had a charge remaining ratio of 52% and a 93% charge remaining ratio was achieved in the co-polymer based system when extrapolated to 10 years. The lower nanocrystal preparation temperature and uniform nanocrystal size and distribution of the di-block co-polymer contribute to the slower charge loss ratio.

CONCLUSIONS

In summary, a novel core-shell nanocrystal MOS memory has been proposed and demonstrated. Wide memory windows of 7.9V and 8.4V was achieved at the ±12 V voltage sweep for memories based on randomly distributed and ordered Co/HfO_2 core-shell nanocrystals, respectively. Uniform particle size and spacing are obtained by a di-block co-polymer fabrication process. The uniform nanocrystal distribution throughout the sample is critical for device reliability and manufacturability. A low-temperature core-shell synthesis process minimizes possible reaction between the metal core and tunneling layer that may deteriorate the device retention performance. The simple and reliable di-block co-polymer process to make core-shell nanocrystal memory may open new opportunities for memory applications.

ACKNOWLEDGMENTS

The authors acknowledge the financial and program support of FCRP center on Function Engineered NanoArchitectonics (FENA), and the National Science Foundation (ECCS-0725630), and the Defense Microelectronics Activity (DMEA) under agreement number H94003-09-2-0901.

REFERENCES

1. S. Tiwari, F. Rana, H. Hanafi, A. Hartstein, E.F. Crabbe and K. Chan, *Appl. Phys. Lett.*, **68** p. 1377, (1996).
2. Y.C. King, T.J. King and C. Hu, *IEEE Trans. Electron Devices*, **48** p. 696, (2001).
3. Y. Shi, K. Saito, H. Ishikuro and T. Hiramoto, *J. Appl. Phys.*, **84** p. 2358, (1998).
4. L.C. Wu, M. Dai, X.F. Huang, W. Li and K.J. Chen, *J. Vac. Sci. Technol.*, B **22**, p. 678, (2004).
5. P.H. Yeh, L.J. Chen, P.T. Liu,D.Y. Wang, and T.C. Chang, *Electrochimica Acta,* **52** pp. 2920–2926, (2007)
6. S. Choi, S. S. Kim, M. Chang, H. S. Hwang, and etc., *Appl. Phys. Lett.*, **86**, 123110 (2005).
7. Yu-Hsien Lin, Chao-Hsin Chien, Ching-Tzung Lin, Chun-Yen Chang, and Tan-Fu Lei, *IEEE Electron Device Letters*, **26**, pp. 154-156 (2005)
8. J. H. Chen, W. J. Yoo, D. S. H. Chan, and L. J. Tang, *Appl. Phys. Lett.*, **86**, 073114 (2005).
9. Eunkyeom Kim, Kyongmin Kim, Daeho Son, Jeongho Kim, Kyungsu Lee, Sunghwan Won, Junghyun Sok, Wan-Shick Hong, and Kyoungwan Park, *Journal of Semiconductor Technology and Science*, **8**, pp. 27-31, (2008)
10. R. Ohba, N. Sugiyama, K. Uchida and etc., *IEEE Trans. Electron Devices*, **49**, 1392 (2002).
11. M. Takata, S. Kondoh, T. Sakaguchi, H. Choi, J. C. Shim, H. Kurino, and M. Koyanagi, *Tech. Dig. - Int. Electron Devices Meet.*, 553, (2003).
12. C. Lee, A. Gorur-Seetharam, and E. C. Kan, *Tech. Dig. - Int. Electron Devices Meet.*, 557, (2003).
13. Huimei Zhou, Reuben Gann, Bei Li, Jianlin Liu and J. A. Yarmoff, *Mater. Res. Soc. Symp. Proc.*, Vol. 1160 1160-H01-05 (2009)
14. Z. Liu, C. Lee, V. Narayanan, G. Pei, and E. C. Kan, *IEEE Trans. Electron Devices*, **49**, 1606, (2002).
15. JooHyung Kim, JungYup Yang, JunSeok Lee, and JinPyo Hong, *Appl. Phys. Lett.*, **92**, 013512, (2008)

16. Jong Jin Lee, Yoshinao Harada, Jung Woo Pyun and Dim-Lee Kwong, *Appl. Phys. Lett.*, **86**, 103505, (2005)
17. J. Robertson, O. Sharia and A. A. Demkov, *Appl. Phys. Lett.*, **91**, 132912 (2007)
18. M. Niwa, *IEDM-SC*, (2000)
19. Yan Zhu, Bei Li, and Jianlin Liu, *J. Appl. Phys.*, **101**, 063702 (2007)
20. Y-C. Perng, J. A. Dorman, S. Gachot, Y. Mao and J. P. Chang, in preparation.

Mater. Res. Soc. Symp. Proc. Vol. 1250 © 2010 Materials Research Society   1250-G06-03

Nanocrystal memory device utilizing GaN quantum dots by RF-MBD

P. Dimitrakis[1], E. Iliopoulos[2, 3], P. Normand[1]
[1]Institute of Microelectronics, NCSR "Demokritos", 15310 Aghia Paraskevi, Greece
[2]Physics Department, University of Crete, P.O.Box 2208, 71003 Heraklion, Greece
[3]Microelectronics Research Group, IESL-FORTH, P.O.Box 1527, 71110 Heraklion, Greece

ABSTRACT

The growth of GaN-QDs by radio frequency plasma assisted molecular beam deposition (RF-MBD) on thin SiO_2 films for non-volatile memories (NVM) applications is demonstrated. Thermal budget modification during the deposition allows tuning of the size and density of the QDs. Preliminary electrical characterization of GaN-QD MOS devices reveals efficient electron injection at very low voltages from the Si accumulation layer to the QDs. The observed limitation in hole injection relates adequately to the energy band diagram of the structure.

INTRODUCTION

The concept of nanocrystal nonvolatile memories (NC-NVM) has been intensively explored in the last decade [1]. Although several issues in terms of uniformity and repeatability of their characteristics are still under investigation, NC-NVMs remain one of the emerging NVM alternatives [2]. Various approaches have been proposed to successfully face the "voltage-time dilemma" [3], i.e. the compromise between high-speed memory programming at low voltage and long data retention time, which is one the main constraints for the adoption of NC-NVMs. Among them the tunnel barrier engineering approach [4] and the use of metal nanoparticles with high work function [5] should be distinguished.

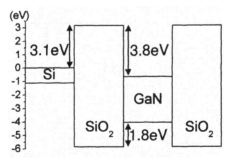

Figure 1. Energy band diagram of the fabricated GaN-QD MOS structure [6].

The work function engineering of the NCs may improve the performance of NVMs in terms of operation voltages and retention time. This can be achieved by forming semiconductor NCs exhibiting negative conduction band offset with respect to the Si substrate conduction band. In such case, the injected electrons can be trapped into the NCs at low voltage while, the retention time can be significantly improved due to the higher energy barrier the carriers have to overcome in order to tunnel back to the Si substrate. Gallium nitride (GaN) NCs or quantum dots

(QDs) fulfill these requirements, as shown in the band-alignment diagram of Figure 1 [6]. In this direction, we focused our efforts on the realization of GaN-QDs structures embedded into SiO_2 layers and on the study of their charge storage properties.

EXPERIMENTAL

High quality 3.5nm thick SiO_2 layers were thermally grown in dry O_2 atmosphere on 1-2 Ωcm n-type 100mm Si (100) wafers. The SiO_2 layer thickness was verified by spectroscopic ellipsometry (SE). GaN-QDs were deposited on the SiO_2/Si substrates by radiofrequency nitrogen plasma assisted molecular beam deposition (RF-MBD) and subsequently buried, with an electron beam evaporated SiO_2 layer, in a different deposition reactor. The post-QDs deposition top oxide layer was also studied by SE measurements on a simultaneously deposited reference substrate and was determined to be 18 nm. Standard Al-gate MOS capacitors were fabricated for electrical measurements at a wafer level. For the sake of comparison, control samples, which were not treated in RF-MBD for the deposition of QDs, were also fabricated Capacitance-to-voltage (C-V) and current-to-voltage (I-V) measurements were carried out at room temperature using a HP4284 Precision LCR meter and a HP4140B pA meter, respectively. Utilizing bi-directional C-V measurements in dark, the stored charge density into QDs can be estimated by the observed hysteresis in the flat-band voltage V_{FB}. Two samples with amorphous/polycrystalline GaN QDs were prepared (samples A and B) by exposure of the SiO_2 surface to the gallium beam, under simultaneous irradiation with an active nitrogen flux produced by the RF-plasma nitrogen source, and subsequent annealing (ripening) of the deposited layer. In both cases deposition took place under nitrogen-rich conditions (the active nitrogen species arrival rate was higher that the corresponding one for Ga atoms). Therefore, the deposited GaN amount was controlled by the Ga flux and the deposition time. For the case of sample A, this was equivalent to 10 monolayers (MLs) of GaN, while for sample B, an equivalent of 6 MLs of GaN was deposited. The deposition was monitored by reflection-high energy electron diffraction (RHEED). While the SiO_2/Si RHEED pattern was completely diffuse, indicating an amorphous surface, the patterns of the deposited GaN layers exhibited broad diffraction rings, indicating that the deposited material was, to a certain degree, of polycrystalline nature. Except the difference in the deposited GaN amount, sample B was annealed at a higher temperature during the ripening stage.

RESULTS AND DISCUSSION

The as-deposited GaN QDs layers were examined by atomic force microscopy (AFM). The corresponding AFM micrographs are shown in Figure 2. As seen in those micrographs, the QDs density for sample B was in the low-10^9 cm^{-2} range while the corresponding one for sample A was approximately 10^{10} cm^{-2}. Furthermore, the average diameter and height were 25nm and 1.7nm for sample A, while sample B exhibited larger average QDs size, i.e. average diameter 44nm and average height 3.5nm. The difference, in the GaN QDs geometrical characteristics, can be attributed to the increased ripening temperature, which resulted in lower surface density and bigger size nanocrystallites, for the case of sample B.

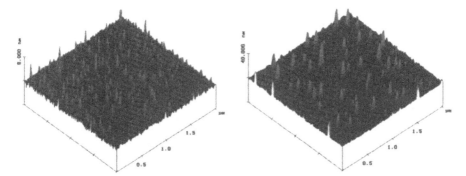

Figure 2. AFM 2×2 μm^2 micrographs of as deposited GaN QDs on SiO$_2$/Si (100) for samples A (left) and B (right). z-ranges of the micrographs are 8 and 40 nm correspondingly.

In Figure 3, the high frequency (1MHz) C-V characteristics obtained in the control sample are presented. No hysteresis was observed up to ±4V that is the maximum voltage that the samples can withstand without degradation. This limited voltage range relates to the poor quality of the control oxide used to cover the GaN QDs. The absence of hysteresis in bidirectional C-V characteristics suggests that the tunnel and control oxides do not suffer from remarkable interface or bulk oxide traps where the injected carriers can be stored. Clearly, the flat-band voltage in these samples is shifted to negative values. This denotes the presence of a high concentration of positive fixed charges probably into the deposited oxide.

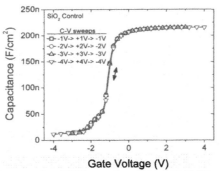

Gate Voltage (V)
Figure 3. Sequential C-V sweeps for a MOS capacitor without GaN-QDs.

The observed stretch-out close to the threshold voltage region (moderate inversion) is mainly attributed to the existence of fast states (D$_{it}$) at the interface between the thin injection oxide and the Si substrate. The total oxide thickness obtained from maximum capacitance (CET) was estimated to be around 16nm, assuming that the dielectric constant for SiO$_2$ was equal to 3.9.

65

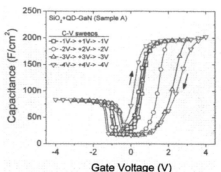

Gate Voltage (V)
Figure 4. Sequential C-V (1MHz) characteristics for a MOS capacitor with small-size GaN-QDs (sample A).

Charge-trapping properties of the fabricated NC MOS capacitors were investigated by C-V measurements. Round voltage sweeps with symmetric limits were performed on different capacitors from each sample. Results from samples A are shown in Figure 4. The C-V branch from inversion to accumulation (forward sweep) is called *hole-injection branch* while the other from accumulation to inversion (backward sweep) is called *electron-injection branch*. Obviously, the hysteresis in C-V characteristics is getting wider as the voltage sweep limits increase. The observed hysteresis can be quantified in terms of V_{FB} shift for each branch, i.e. $\Delta V_{FB}=V_{FB+}-V_{FB-}$, where V_{FB+} and V_{FB-} are the flat-band voltages of the electron-injection and hole-injection branches, respectively [7]. It is clear that the V_{FB-} values remains almost constant with increasing of the applied voltage sweeps suggesting that hole-injection is a limited process. This is expected because Si substrate holes have to overcome a high energy barrier in order to be injected into the QDs, as shown in Figure 1.

On the contrary, electrons are easily injected and trapped into the GaN-QDs and the V_{FB+} values increase with the voltage sweeps. A similar behavior was also detected for sample B as shown in Fig. 5. The only difference between samples A and B charging properties relates to the somewhat higher V_{FB-} values and thereby, the larger memory window (ΔV_{FB}) attainable in sample A (Fig. 5). This effect is mainly attributed to the limited efficiency of sample B to extract the trapped electrons during the voltage sweep from inversion to accumulation.

In Fig. 6, the C-V characteristics of the examined samples are compared. Clearly, the control sample is suffering from a large density of positive fixed charges, which must be attributed to the e-gun deposited control SiO_2 layer at room temperature. Samples A and B did not exhibit such a behavior and exhibit very similar V_{FB} values under discharge conditions. The latter observation indicates that the different MBD deposition conditions of the GaN QDs did not affect the fresh V_{FB} value of the MOS capacitors.

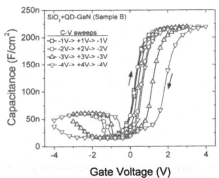
Figure 5. Sequential C-V (1MHz) characteristics for MOS capacitor with small-size GaN-QDs embedded in gate stack (sample B).

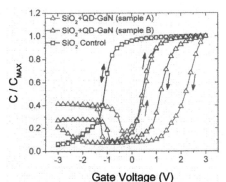
Figure 6. Comparative normalized C-V (1MHz) plots for control and two samples with embedded GaN QDs.

Finally, we compared the equivalent parallel conductance measurements, G-V, obtained simultaneously with the C-V characteristics shown in Figure 6. For all samples we can distinguish the presence of two distinct peaks indexed by "1" and "2". A careful examination of the A1f and A1b conductance peaks, obtained after forward and backward voltage sweeps respectively, reveals a peak-maximum voltage difference equal to the ΔV_{FB} extracted from the C-V characteristics. A similar situation occurs in sample B for peak B1. The full width at half maximum of the A1 and B1 conductance peaks are similar. These observations indicate that the V_{FB} hysteresis and the conductance peak have the same origin. Furthermore, the heights of the conductance peaks after forward and backward voltage sweeps are the same in both samples A and B suggesting that the electron discharging and charging effects have the same origin too. For the control sample, the peaks C1f and C1b occur at the same voltage which is consistent with the absence of hysteresis in the C-V characteristics. From the above experimental results we can

67

conclude that the observed C-V hystereses and voltage shifts of the A1/B1 conductance peaks are due to charge trapping/detrapping processes taking place in GaN QDs and not in defects located at the Si substrate/SiO$_2$ interface. A clear signature of the latter defects, commonly referred to as interface traps, is given by the A2, B2 and C2 conductance peaks.

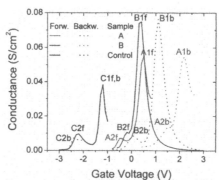

Figure 7. Parallel conductance measurements (1MHz) after series resistance effect correction obtained for the tested MOS capacitors with and without GaN-QDs.

CONCLUSIONS

Pyramidal GaN QDs were successfully fabricated by RF-MBD on 3.5nm SiO$_2$/Si substrates. By varying the parameters of the GaN deposition, different QD distributions can be achieved. The QDs were integrated in Si MOS capacitors exhibiting significant electron trapping. C-V and G-V measurements revealed that this is due electron charging/discharging of GaN QDs.

REFERENCES

1. P. Dimitrakis, P. Normand, D. Tsoukalas, "Si Nanocrystal Memories", in *Silicon Nanophotonics*, ed. L. Khriachtchev, Pan Publishing, Singapore, 2008 (ISBN-13 978-981-4241-11-3).
2. S. K. Lai, IBM J. Res. Dev. **52**, 529 (2008).
3. T.-H. Hou, C. Lee, V. Narayanan, U. Ganguly, E.C. Kan, IEEE Trans. Elect. Dev. **53**, 3095 (2006).
4. S.J. Baik, S. Choi, U-In Chung, J.T. Moon, Solid-State Electron. **48**, 1475 (2004.)
5. H. Schroeder, V. V. Zhirnov, R. K. Cavi, R. Waser, J. Appl. Phys. **107**, 054517 (2010).
6. Kyong-Hee Joo, Chang-Rok Moon, Sung-Nam Lee, Xiofeng Wang, Jun Kyu Yang, In-Seok Yeo, Duckhyung Lee, Okhyun Nam, U-In Chung, Joo Tae Moon, Byung-Il Ryu, IEDM Tech. Dig. 2006, doi:10.1109/IEDM.2006.346950.
7. P. Dimitrakis, E. Kapetanakis, P. Normand, D. Skarlatos, D. Tsoukalas, K. Beltsios, A. Claverie, G. Benassayag, C. Bonafos, D. Chassaing, V. Soncini, Material Science & Engineering B **101**, 14-18 (2003).

Mater. Res. Soc. Symp. Proc. Vol. 1250 © 2010 Materials Research Society   1250-G06-04

Formation of Ge Nanocrystals in High-k Dielectric Layers for Memory Applications

P. Dimitrakis[1], V. Ioannou-Sougleridis[1], P. Normand[1], C. Bonafos[2], S. Schamm-Chardon[2], A. Mouti[2], B. Schmidt[3], J. Becker[4]
[1]Institute of Microelectronics, NCSR "Demokritos", 15310 Aghia Paraskevi, Greece
[2]CEMES-CNRS, Universite de Toulouse, nMat group, BP 94347, 31055 Toulouse cedex 4, France
[3]Research Center Dresden-Rossendorf, Dresden, Germany
[4]Cambridge Nanotech Inc, Cambridge, MA, USA

ABSTRACT

This paper reports on the fabrication of Ge-NCs in Al_2O_3 and HfO_2 layers by ion-beam-synthesis for memory applications. After furnace annealing at 800°C, Ge-NCs form in Al_2O_3 materials as revealed by TEM and EELS investigations, while no signature of such NCs is observed in the HfO_2 layers. The charge storage properties of the non- and Ge-implanted Al_2O_3 layers were examined as a function of the annealing temperature in the 700-1050°C range using MIS capacitors. Strong charge storage is detected in the 800-950°C-annealed and Ge-implanted Al_2O_3 layers leading to large and similar memory windows. The I-V characteristics of SiO_2-capped Ge-implanted-HfO_2 structures exhibit significant negative-differential-resistance effects probably due to the formation of conductive paths made of hafnium germanide ($HfGe_2$) or hafnium germanate (HfGeO) regions.

INTRODUCTION

Semiconductor nanocrystals (NCs) have been employed successfully in various demonstrators for new CMOS device applications like nonvolatile memory (NVM) cells [1] and optoelectronic components [2]. Ge nanocrystals are of special interest for nano-floating gate NVM cells due to their band offsets with respect to the bands of the Si substrate; an attractive NC property for faster programming speeds and longer retention times compared to silicon NCs [3]. Another interesting option is the use of high-k materials as tunneling and blocking dielectrics. High-k materials offer the advantages (1) to prevent the back tunneling of trapped charges through the use of thick tunneling layers and (2) to increase the coupling coefficient between the control gate and the nano-floating gate when they are employed as blocking dielectrics [4]. Furthermore, at retention mode the internal electric caused by trapped charges is significantly lower precluding the charge loss due to tunneling through the injection and control oxide layers. This route has been examined to overcome the charge retention issues rising from the thinning of the injection SiO_2 layer. Based on the above, Ge-NCs embedded in high-k dielectrics provide a promising alternative in the development of NC-memories. This work examines the prospect of using the low-energy ion-beam-synthesis (LE-IBS) technique for fabricating Ge-nanocrystals in thin Al_2O_3 and HfO_2 layers. This CMOS compatible and time-proven technique has been successfully employed the last ten years in the formation of Si [5] and Ge [3] nanocrystals in thin SiO_2 films. Structural and electrical characterization of non-implanted and Ge+-implanted Al_2O_3 and HfO_2 layers are herein reported and discussed in view of the development of NVM cells.

EXPERIMENTAL

7nm-thick Al_2O_3 layers were formed on n-Si substrates by atomic layer deposition (ALD) and subsequently subjected to 1keV Ge^+ implantation at a dose of $1\times10^{16}cm^{-2}$. Next, the implanted layers were covered with a 10nm-thick ALD Al_2O_3 layer and furnace annealed for 20min at temperatures in the 700-1050°C range in N_2 ambient. Non-implanted control samples were also fabricated and annealed under the same conditions. The electrical and memory characteristics of the prepared structures were studied by I-V and C-V-f measurements utilizing Al-gate MIS capacitors fabricated by conventional Al deposition, photolithography and Al etching techniques. The final structure and fabrication conditions of the tested samples are shown in Fig.1(a) and Table I respectively. 7nm-thick ALD-prepared HfO_2 layers were 1keV Ge^+ implanted at doses of 0.5 and $1\times10^{16}cm^{-2}$. The as-implanted samples were covered by 9nm-thick SiO_2 using TEOS LPCVD at 710°C following by in situ densification at 800°C for 20min. In this case, post-implantation annealing (PIA) occurred during the deposition and densification steps of the SiO_2 capping layer. The final structures were electrically characterized using Al-gate MIS capacitors. Specimens transparent to electrons were prepared for cross-sectional transmission-electron-microscopy (XTEM) by using the standard procedure. High-resolution (HR) TEM and electron-energy-loss spectroscopy (EELS) analyses were performed on a field emission TEM, FEI Tecnai™ F20 microscope operating at 200 kV, equipped with a corrector for spherical aberration.

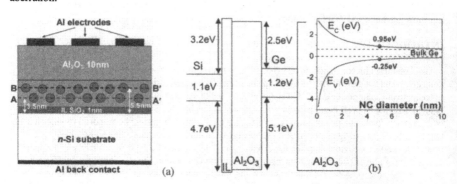

Figure 1. (a) Final MIS structure with embedded Ge-NCs after annealing. The lines AA' and BB' denote the average distances from Si substrate to the centers of NCs as extracted by TEM statistics [6]. (b) Energy band diagram of the tested Al_2O_3 structures. Inset shows the evolution of the bandgap with Ge-NCs size as calculated in [7].

DISCUSSION

TEM studies

Figure 2 shows an XTEM image of a Ge-implanted and 800°C-annealed Al_2O_3 sample. Obviously, an interfacial layer (IL) between the high-k layer and the Si substrate is formed. The implanted and the capping high-k layers are crystalline (γ-cubic phase) and spherical crystalline regions with darker contrast can be distinguished. EELS analysis revealed that the IL is made of SiO_2 and the observed spherical crystalline regions are Ge-NCs. Quantitative XTEM analyses [6] revealed also that the NCs have an average size of 5nm and their centroids are distributed along two different lines (Fig.1a). The first locus of centroids is at a distance of 3.5nm from the Si substrate, i.e. the outer surface of the nanocrystals is in touch with the IL. The second locus of NCs is arranged at a distance of 5.5nm from the Si substrate.

Table I. Fabrication conditions of the examined MIS capacitors with Ge-ion implanted 7nm Al_2O_3 gate insulator and 10nm Al_2O_3 capping layers.

Furnace Annealing Temperature	Control Samples	Implanted samples (1keV 1×10^{16} Ge$^+$cm^{-2})
700°C	R4R14	H4124
800°C	R4R12	H4122
950°C	R4R15	H4125
1050°C		H4126

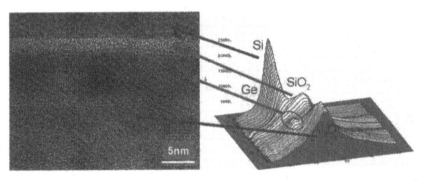

Figure 2. XHREM and EELS line spectrum of an Al_2O_3 thin film implanted with 10^{16} Ge$^+$ cm^{-2} and annealed at 800°C.

No NCs were observed in the SiO_2-capped Ge-implanted-HfO_2 layers. Figure 3 shows that a 3.5nm-thick SiO_2 layer (IL) was formed between the Si substrate and the HfO_2 layer Further, TEM observations (not presented here) revealed that the implanted HfO_2 layer was crystalline after TEOS deposition. Preliminary XRD results for high-dose implanted samples (not shown here) suggest the presence of tetragonal Hf_3GeO_8 and orthorhombic HfO_2. XRD studies on this subject are currently in progress.

Electrical characterization results

Figure 1(b) represents the energy band diagram of the implanted Al_2O_3 structures. The Ge-NCs bandgap was calculated to be 1.2eV according to the theoretical studies by Niquet et al.

[7]. The inset shows the evolution of the minimum of the conduction (E_C) and maximum of the valence (E_V) bands as a function of the Ge-NCs diameter. The changes in E_C and E_V are almost similar resulting to similar band-offsets with the Si bandgap edges, i.e. 0.3eV and 0.25eV between the conduction and valence bands for 5nm NCs, respectively. According to the above energy band diagram, hole injection to Ge-NCs appears as a very efficient process under inversion conditions while the band-offset difference limits hole back-tunneling. In the latter case, the holes have to overcome a higher energy barrier and thereby, they are expected to exhibit larger retention times than electrons.

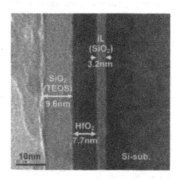

Figure 3. Defocused bright field XTEM image of Ge-implanted and SiO_2-capped (TEOS) HfO_2 layer structure.

Charge storage characteristics of the non- and Ge-implanted Al_2O_3 samples described in Table I were investigated by I-V and C-V measurements of MIS capacitors through the application of successive and symmetric round voltage sweeps starting from inversion. I-V characteristics of non-implanted samples (not shown here) revealed that the leakage current is reduced with increasing annealing temperature, i.e. with increasing films crystallinity, suggesting that crystalline Al_2O_3 has improved insulating properties compared to amorphous. This is mainly attributed to the reduction of defects and to the increase of bandgap. All C-V characteristics [6] exhibit clockwise hysteresis due to the charge storage taking place during the voltage sweep; electron (hole) injection occurs when the device is biased in inversion (accumulation). The hysteresis constitutes the memory window of the device and is extracted from the difference between the flat-band voltage V_{FB} shifts (ΔV_{FB}) measured in the inversion and accumulation regimes of each bi-directional C-V curve. ΔV_{FB} is calculated with respect to the V_{FB} value of a pristine capacitor (i.e., for stressing voltages below 2V). ΔV_{FB} versus the applied sweep voltage characteristics for the implanted samples are shown in Figure 4(a) and are compared to those extracted from the control samples in figure 4(b).

Strong charge storage effects are detected for the Ge-implanted Al_2O_3 samples annealed at 800 and 950°C. The V_{FB} shifts induced by electron or hole storage in these two samples increase with the applied voltage with a trend to saturation and lead to similar memory windows. Above 950°C, trapping of electron and hole is effective but the memory windows decrease significantly probably due to the loss of Ge via the formation of volatile GeO or GeH_4 species. The formation of volatile Ge compounds and subsequent Ge loss during annealing is attributed to the presence of oxygen and hydrogen atoms originating from the air humidity and wet cleaning of the samples [8]. The sample treated at 700°C exhibits only electron trapping and their partial detrapping at negative voltages. The same behavior is observed for all control samples after annealing as shown in figure 4(b). This is an indication that the presence of Ge-NCs is playing a critical role in electron/hole trapping. For the sake of comparison, in figure 4(b) the memory window of Ge-implanted samples subjected to PIA at 1050°C is added. Although, for control samples the maximum V_{FB} shift due to electron trapping is similar to that of implanted sample annealed at 1050°C, the latter exhibits hole trapping, which is not observed in the characteristics of the control samples.

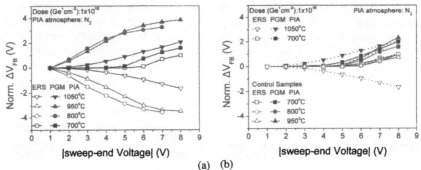

(a) (b)

Figure 4. Memory windows obtained from C-V hysteresis using MIS capacitors described in Table I: (a) Ge-implanted samples and (b) control samples. ERS and PGM denote the erase and program states respectively.

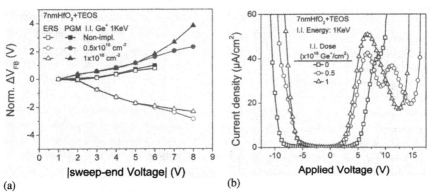

(a) (b)

Figure 5. (a) Memory windows obtained from C-V hysteresis and (b) I-V characteristics under forward and reverse bias obtained using MIS capacitors fabricated using Ge-implanted HfO₂ with TEOS SiO₂ as control oxide.

The flat-band voltage shifts extracted from C-V measurements of the SiO_2-capped non- and Ge-implanted HfO_2 samples are shown in figure 5(a). Similar to the Al_2O_3 samples, hole injection is not observed in the control samples. For the implanted samples, electron charge storage increases with the implantation dose at high sweep voltages (> 6V), while hole charging remains nearly the same. According to TEM analysis, no Ge-NCs were observed suggesting that the measured electron/hole storage into HfO_2 is probably due to the charging of bulk oxide defects and grain boundaries of the various crystalline structures/phases of the HfGeO and HfO_2 materials.

The I-V characteristics of the high-dose implanted samples exhibit enhanced conduction detected through the gate stack at low electric fields due to the presence of grain boundaries between the Hf_3GeO_8 and HfO_2 crystalline phases. The later has slightly lower bandgap compared to the former, i.e. there are non-negligible band-offsets. Electron traps located in the boundary regions allow for the trap-assisted tunelling conduction. When the traps are filled the electrons are blocked by the energy barriers at the boundaries causing the reduction of the current flow and a negative differential resistance (NDR) effect is rising in a specific reverse bias voltage regime.

CONCLUSIONS

The structural and electrical properties of non-implanted and Ge-implanted 7nm Al_2O_3 and HfO_2 oxides after thermal annealing were investigated. TEM and EELS studies revealed the formation of spherical Ge-NCs in the case of implanted Al_2O_3 layers after annealing at a temperature of 800°C while no NCs were observed in the implanted HfO_2 layers annealed at the same temperature. C-V measurements performed on MIS capacitors reveal strong charging effects for the Ge-implanted Al_2O_3 samples annealed in the 800-950°C range resulting from electron and hole storage into the NCs. In that case the attainable memory windows present few differences, thus indicating the formation of stable structures for a large annealing temperature range. TEM studies in implanted HfO_2 layers with TEOS SiO_2 as control oxide revealed the formation of crystalline hafnium gernamate or germanide regions. These crystalline phases may contribute to the existence of NDR regions observed in the I-V characteristics of the related MIS capacitor structures.

REFERENCES

1. S Lai, IBM J. Res. Dev. 52 (2008) 529.
2. L Khriachtchev, *Silicon Nanophotonics*, World Scientific Publishers (2008)
3. V. Beyer, J. von Borany, M. Klimenkov, J. Appl. Phys. 101, 094507 (2007)
4. X.B. Lu, P.F. Lee, J.Y Dai, Thin Solid Films 513, 182 (2006)
5. C. Bonafos, M. Carrada, N. Cherkashin, H. Coffin, D. Chassaing, G. Ben Assayag, A. Claverie, T. Müller, K. H. Heinig, M. Perego, M. Fanciulli, P. Dimitrakis, and P. Normand, J. Appl. Phys. 95, 5696 (2004)
6. P. Dimitrakis, A. Mouti, C. Bonafos, S. Schamm, G. Ben Assayag, V. Ioannou-Sougleridis, B. Schmidt, J. Becker, P. Normand, Microelectr. Engineer. 86, 1838 (2009)
7. Y. M. Niquet, G. Allan, C. Delerue and M. Lannoo, Appl. Phys. Lett. 77, 1182 (2002)
8. V. Beyer, J. von Borany, Phys. Rev. B 77, 014107 (2008)

Mater. Res. Soc. Symp. Proc. Vol. 1250 © 2010 Materials Research Society   1250-G06-06

CVD Growth and Passivation of W and TiN Nanocrystals for Non-Volatile Memory Applications

G.Gay[1], D.Belhachemi[1], J.P. Colonna[1], S. Minoret[1], A. Beaurain[2], B. Pelissier[2], M.C. Roure[1], D. Lafond[1], E. Jalaguier[1], G. Molas[1], T. Baron[2] and B. De Salvo[1]
[1]CEA LETI MINATEC, 17 rue des Martyrs, 38054 Grenoble, France
[2]CNRS-LTM, 17 rue des Martyrs, 38054 Grenoble, France

ABSTRACT

In this paper, we present CVD (Chemical Vapor Deposition) growth and passivation of tungsten (W) and titanium nitride (TiN) nanocrystals (NCs) on silicon dioxide and silicon nitride for use as charge trapping layer in floating gate memory devices. NCs are deposited in an 8 inches industrial CVD Centura tool. W and TiN are chosen for being compatible with MOSFET memory fabrication process. For protecting NCs from oxidation, a silicon shell is selectively deposited on them. Moreover, for a better passivation, TiN NCs are encapsulated in silicon nitride (Si_3N_4) in order to get rid of oxidation issues. After high temperature annealing (1050°C under N_2 during 1 minute) XPS measurements point out that NCs are still metallic, which makes them good candidates for being used as charge trapping layer in floating gate memories.

INTRODUCTION

Thanks to their discrete nature and intrinsic robustness towards defects in the surrounding dielectrics, Silicon nanocrystal (Si-nc) trapping layers offer several advantages on standard poly-Si floating gates, as improved data retention after endurance in particular at high temperatures[1,2]. It has also been shown that coupling the Si-nc concept with high-k control dielectrics, by improving the gate coupling ratio, enables Fowler-Nordheim (FN) program/erase, thus opening the paths for NAND Flash application[3]. However, one of the key limitations of Si-nc memories is the limited memory window which is not suitable for multi-level memory applications. Metal nanocrystals are candidates to increase the number of charges stored (larger density of states and larger work function). One way to obtain a dense array of metal nanocrystals is to evaporate and a thin metal layer followed by a rapid thermal annealing[4,5]. However, in the frame of 3D integration, a conformal depositon of microelectronics metal nanocrystals is necessary. In this paper, we present the CVD growth of tungsten and titanium nitride nanocrystals. Passivation of these nanocrystals by a silicon shell is also presented.

EXPERIMENT

Nanocrystals growth and passivation are done in an 8 inches industrial CVD Centura tool. All fabrication steps are made in different chamber but in the same equipment during the same run. XPS measurements were performed on a customized Thermo Electron Theta 300 spectrometer, directly interfaced to the Alcatel vacuum carrier pod via a vacuum transfer chamber (10 mbar range). All measurements were performed on full 200 mm wafers without any cut. A Hitachi MEB5000 scanning electron microscope (SEM) is used for nanocrystals observation. High resolution transmission electron microscope (HRTEM) analysis are performed to confirm crystallinity of the nanocrystals

DISCUSSION

Growth of Tungsten nanocrystals

The deposition is done in two steps, at 440°C. The first step is a silane (SiH_4) soak during which silicon oxide surface is exposed to a SiH_4 flux. In a second step, tungsten precursor WF_6 is flowed in the reactor and reduced by SiH_4 to form W nuclei on the surface. Fluorine is extracted in the gas phase SiF_4.

$$2WF_{6 (g)} + 3\ SiH_{4 (g)} => 2W + 3\ SiF_{4 (g)} + 6H_{2 (g)}$$

SEM and HRTEM observation of nanocrystals are shown on Figure 1. The nanocrystals are well-separated and crystallinity is clearly confirmed on HRTEM image. A 7.10^{11} cm^{-2} nanocrystals density is extracted on the SEM images.

Figure 1: (a) SEM observation (40° tilt) of W nanocrystals **(b)** planar HRTEM observation of W nanocrystals

Growth of Titanium Nitride nanocrystals

Titanium nitride (TiN) nanocrystals are deposited at 680°C with CVD precursors $TiCl_4$ and NH_3 CVD reaction is described in the next equation:

$$6TiCl_4\ (g) + 8NH_3\ (g) => 6TiN(s) + 24HCl\ (g) + N_2\ (g)$$

SEM observations allow for extracting NCs density. TiN nanocrystals density is larger than 10^{12} cm^{-2}. However, they are still physically separated as observed on HRTEM image on Figure 2. HRTEM images also prove the crystallinity of the NCs.

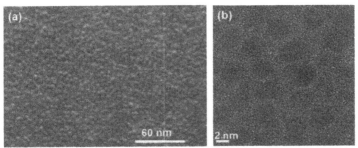

Figure 2: (a) SEM observation (40° tilt) of TiN nanocrystals (b) planar HRTEM observation of TiN nanocrystals

Passivation of deposited nanocrystals

Passivation is done by depositing a thin silicon layer on silicon nanocrystals. The silicon source used is dichlorosilane DCS SiH_2Cl_2 gas. Thanks to this gas, silicon is selectively deposited on metal surface at 550°C. No deposition occurs on the silicon oxide at this temperature. In order to highlight the passivation efficiency, the nanocrystals are exposed to a SC1 (standard cleaning) etching. SC1 etching (composed of NH_4OH, H_2O_2 and H_2O) is etching TiN ten times faster than silicon and 25 times faster than silicon oxide. We observe on Figure 3 that the SC1 attack has no effect on passivated TiN nanocrystals while it has etched totally the non passivated nanocrystals. This means that a silicon shell has been deposited on top of the TiN nanocrystals, protecting them from SC1 etching. Moreover, this additional silicon layer will act as an additional trapping layer.

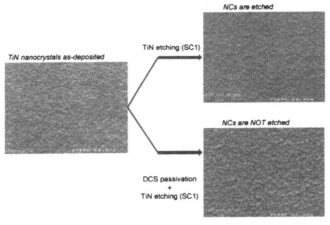

Figure 3: Highlight of the DCS passivation. TiN nanocrystals are not etched in SC1 solution when they are passivated by DCS because of a silicon shell protection.

77

XPS measurements on nanocrystals

- **W nanocrystals**

In order to study the chemical nature of the metal nanocrystals, XPS measurements are performed on the nanocrystals. On Figure 4, it appears that without passivation, tungsten is mainly in an oxide form, either WO3 or WO2. Indeed, it is well-known that W oxidizes at ambient[6]. W is first oxidized in WO2 form and in a second step attains the WO3 form. On contrary, silicon passivated W NCs are very few oxidized after exposure to ambient. This is explained by the fact that silicon oxidation is thermodynamically favourable compared to W oxidation[1]. As a consequence, silicon is an oxygen getter protecting W from oxidation.

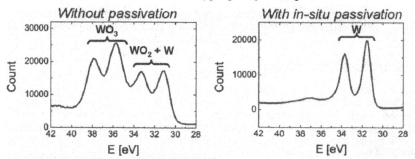

Figure 4: XPS measurements of passivated and non passivated W nanocrystals on SiO2 after exposure to ambient

Since W nanocrystals are dedicated to be used as trapping layer in floating gate memory, we have tested the thermal stability of the passivated nanocrystals. A 2 nm thick Al$_2$O$_3$ layer is deposited by atomic layer deposition (ALD) on top of the nanocrystals in order to simulate the blocking oxide of a memory gate stack. Then the highest temperature of our MOS thermal budget is applied to the stack (1050°C during one minute in N$_2$). XPS measurements are then made to measure the chemical nature of tungsten. Results are summarized on Figure 5. We observe that more than 70% of the nanocrystals are under metallic W form. The partially oxidized nanocrystal structure can be interpreted as a thin tungsten oxide shell around the metallic tungsten nanocrystal core. This is a proof that DCS passivation is robust enough to endure high thermal budget.

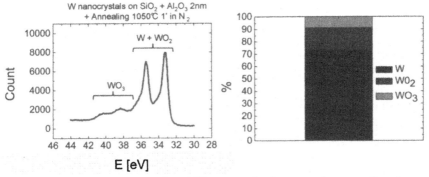

Figure 5: XPS measurement on passivated W nanocrystals after a one minute annealing at 1050°C in nitrogen.

- **TiN nanocrystals**

Titanium nitride is more sensitive towards oxidation. As a consequence, DCS passivation alone is not enough to protect TiN from oxidation. A pure TiO_2 nanocrystal is obtained in that case. The solution is to insulate TiN nanocrystals from oxygen atoms. This is done by first depositing nanocrystals on a thin nitride layer (Si_3N_4), then passivating the nanocrystals with DCS, and finally encapsulating the passivated nanocrystals in a second thin nitride layer. In this way, nanocrystals are protected from oxygen since nitride is known for being a good oxygen diffusion barrier. XPS measurements plotted on Figure 6 show that after a thermal annealing of 1 minute at 1050°C in nitrogen, nanocrystal is composed of a metallic TiN core with a TiO_2 shell surrounding it.

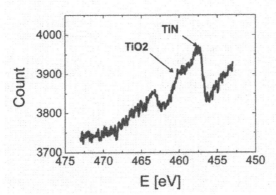

Figure 6: Ti XPS spectra of DCS passivated TiN nanocrystals encapsulated in SiN (after a 1050°C, 1 minute thermal annealing in N_2)

CONCLUSIONS

This work reports on the growth of TiN and W nanocrystals. High density, crystalline nanocrystals arrays are obtained. An in-situ passivation process is also presented, protecting metal nanocrystals from oxidation at ambient. This process is based on a selective deposition of silicon on the nanocrystals. At 1050°C, silicon passivation alone is enough to protect W nanocrystals from oxidation. On the contrary, TiN nanocrystals must be encapsulated in a nitride layer to be able to resist at this thermal budget. These metal nanocrystals processes are thus suitable for integration as a trapping layer in flash memory devices.

REFERENCES

1. G. Chindalore, J. Yater, H. Gasquet, M. Suhail, S. Kang, C. M. Hong, N. Ellis, G. Rinkenberger, J. Shen, M. Herrick, W. Malloch, R. Syzdek, K. Baker, K. Chang, "Embedded split-gate flash memory with silicon nanocrystals for 90nm and beyond" VLSI Tech. Dig., pp.136-137, 2008.

2. C. Gerardi, G. Molas, G. Albini, E. Tripiciano, M. Gely, A. Emmi, O. Fiore, E. Nowak, D. Mello, M. Vecchio, L. Masarotto, R. Portoghese, B. De Salvo, S. Deleonibus, A. Maurelli, "Performance and reliability of a 4Mb Si nanocrystal NOR Flash memory with optimized 1T memory cells" IEEE Tech. Dig. of IEDM, pp.821-824, 2008

3. G. Molas, M. Bocquet, J. Buckley, J. P. Colonna, L. Masarotto, H. Grampeix, F. Martin, V. Vidal, A. Toffoli, P. Brianceau, L. Vermande, P. Scheiblin, M. Gely, A. M. Papon, G. Auvert, L. Perniola, C. Licitra, T. Veyron, N. Rochat, C. Bongiorno, S. Lombardo, B. De Salvo, and S. Deleonibus, "Thorough investigation of Si-nanocrystal memories with high-k interpoly dielectrics for sub-45nm node Flash NAND applications" IEEE Tech. Dig. of IEDM, pp.453-456, 2007.

4. S. K. Samanta, Zerlinda Y. L. Tan, Won Jong Yoo, Ganesh Samudra, and Sungjoo Lee, "Self-assembled tungsten nanocrystals in high-k dielectric for nonvolatile memory application" J. Vac. Sci. Technol. B Volume 23, Issue 6, pp. 2278-2283 (November 2005)

5. J. Dufourcq, S. Bodnar, G. Gay, D. Lafond, P. Mur, G. Molas, J. P. Nieto, L. Vandroux, L. Jodin, F. Gustavo and Th. Baron "High density platinum nanocrystals for non-volatile memory applications", Appl. Phys. Lett. 92, 073102 (2008)

6. B. Pelissier, A. Beaurain, J.P. Barnes, R. Gassilloud, F. Martin and O. Joubert, "Parallel angle resolved XPS investigations on 12 in. wafers for the study of W and WSi_x oxidation in air", B. Pelissier et al., Microelectron. Eng. Volume 85, Issue 9, September 2008, Pages 1882-1887

Mater. Res. Soc. Symp. Proc. Vol. 1250 © 2010 Materials Research Society   1250-G06-12

The Quantitative Study of Trapped Charges in Nano-Scale Ge Island by EFM Measurement

Zhen Lin[1], Pavel Brunkov[2], Franck Bassani[3], Georges Bremond[1]

[1] Institut des Nanotechnologies de Lyon, UMR 5270 Université de Lyon, Institut National des Sciences Appliquées, INSA Lyon, Bât. Blaise Pascal, 20, avenue Albert Einstein - 69621 Villeurbanne Cedex, France
[2] IOFFE Physical-Technical Institute of the Russian Academy of Sciences, Saint-Pétersbourg, 194021, Russia
[3] Institut Matériaux Microélectronique Nanosciences de Provence, UMR CNRS 6242, Avenue Escadrille Normandie-Niemen - Case 142, F-13397 Marseille Cedex 20, France

ABSTRACT

In this work, an individual Ge island on top of silicon dioxide layer has been charged by a conductive EFM tip and quantitatively characterized at room temperature. Electrons or holes were successfully injected and were trapped homogenously in the isolated nano-scale Ge island. In order to quantitatively study these trapped charges, a truncated capacitor model was used to approximate the real capacitance between the tip and island surface. The analytical expression of the quantity of trapped charges in isolated Ge island as a function of the EFM phase signal was deduced. Applying a tip bias for -7V during 30 seconds leads to an injection about 800 electrons inside an individual Ge island.

INTRODUCTION

Injection and detection of localized charges in nanostructures, on or below the surface, are the key issues in the field of nanoelectronics. Electrostatic force microscopy (EFM) is proved to be one of the most efficient tools not only for atomic scale characterisation of surfaces[1-2] but also for local surface modification and structuring in sub-micrometre and nanometre ranges[3-5]. Using a conductive tip, it is possible to electrically bias the oscillating tip with respect to the sample and characterize materials for accurate local and non-destructive electrical properties for a wide range of characterisations such as surface potential, charge distribution, doping concentration and dielectric constant[6-9]. It is also possible to use the two-pass lift mode to minimize the influence of morphology over the sample surface by keeping the long range electrostatic force distance constant.

This ability has been used to quantitatively study of the charge injection and retention in some kind of samples[10-12] such as Germanium nano islands deposited directly on oxide layer for non-volatile memory application at room temperature. However, a quantitatively study of these trapped charges inside the isolated nano-scale Ge island is of great importance for the memory applications and is required.

EXPERIMENT DETAILS

These nano-scale Ge islands have been fabricated on a very thin wafer. A 15nm thickness Ge layer was deposited over the 5nm SiO_2 in ambient temperature and was thermal annealed by 750°C for 20 minutes. The Ge nano crystals were formed with an average diameter around 150nm. It is about 7 times the thickness of the Ge layer[24]. Over the dots there is a natural oxide layer about only 2 nm so that the sample surface is uneven.

The EFM two-pass lift mode was used to solve the topographical influence during charge detection. The measurement was conducted on a Veeco Metrology Digital Instruments 3100 Dimensions AFM employing a Nanoscope V controller. Charges were injected by using commercial conductive SCM-PIT tip. Its characters are: elasticity coefficient: 2.8N/m, resonance frequency: 75 kHz and Pt/Ir metal coating.

An individual nano-scale Ge island was chosen for the charge experiment. During the charging process, -7V was applied to the conductive tip during 30 seconds. Then the tip lifted up to 50 nm and polarised (V_{EFM}) by +2V for electrostatic force measurement. Fig.1 shows the capture images after charge injection. These isolated nano-scale Ge islands were 150 nm in average diameter. The phase shift was detected at -4.5° (V_{EFM} = +2V). This means that there were many electrons were injected and trapped inside the isolated Ge island which greatly changed the electrostatic force gradient around this area.

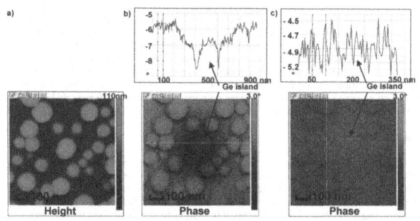

Fig.1 Images after charge injection by -7 V, 30s. Electrons were injected into the isolated Ge island. (a) topography of sample surface; (b) phase signal images, V_{EFM} = +2V, profile on top; (c) phase signal images, V_{EFM} = 0V, profile on top.

THEORY AND DISCUSSION

In order to quantitatively characterize the trapped charges, equivalent structures should be used for modelling the interaction between AFM tip and sample surface.

In EFM measurement, these changes due to the trapped charges could easily be detected by the phase shift $\Delta\phi$[7-14].

$$\Delta\phi = -\frac{Q}{k}\frac{\partial F_s}{\partial z}(z_0) = -\frac{Q}{2k}\frac{d^2C}{dz^2}(V_{EFM} - V_q)^2 \quad (1)$$

Where Q is the quality factor of AFM system, k is the elasticity coefficient, z is the height of the tip, z_0 is the distance between tip and sample, $\partial F_s/\partial z$ is the spatial derivative of the electrostatic force acting on the EFM probe by these charges, V_{EFM} is the tip potential during the EFM measurement.

The AFM probe is composed of three main parts: the cantilever, the tip and the tip apex. During the AFM measurement, these three parts interact with the sample surface and can be modelled simply as a series of flat plane capacitances. The cantilever is modelled as a flat plane, the tip is modelled as a truncated cone[13] and the tip apex is modelled as a small sphere or the tip and its apex as a cone[14]. From the theoretical studies we can know that the major contribution to the capacitance variation is given by tip apex, followed far below by the cone, chip and cantilever. So the truncated cone-plane model is relatively the most accurate one which considers the tip apex and the tip body.

Truncated cone-plane model (total tip apex and tip body) gives the second derivative capacitance factor $C''_{tcp}(z)$:

$$C''_{tcp}(z) \approx 2\pi\varepsilon_0 \left[\frac{R^2(2z+R)}{z^2(z+R)^2} + k^2\left(\frac{1}{z+R} + \frac{R}{\sin\theta(z+R)^2} \right) \right]$$

$$with \quad k^2 = \frac{1}{\ln^2\tan\left(\dfrac{\theta}{2}\right)} \tag{2}$$

Where R is the tip apex radius and θ is the tip-opening angle, and k is the same as that in Eq. (1).

The equivalent electric circuit of this experiment configuration can be presented in Fig.2.

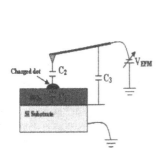

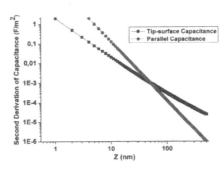

Fig. 2 Equivalent circuit   Fig.3 Second derivative capacitance versus tip surface distance

Where C_1 is the charged island-substrate capacitance, C_2 is the charged island-tip capacitance and C_3 is the substrate-probe capacitance.

Thus, electrostatic force due to these trapped charges could be expressed as:

$$F = \frac{1}{2}\left(\frac{C_1V - q}{C_1 + C_2} \right)^2 \frac{dC_2}{dz} + \frac{1}{2}V^2\frac{dC_3}{dz} \tag{3}$$

For SCM-PIT, the tip height is 10~15um and R is 20nm with tip-opening angle of 10°. The lift height is always set at 50nm. The variation of the second derivative capacitance $C''(z)$ varies with the tip-surface distance z is plotted in fig.3. We can observe that the effective surface during the tip scan is about 7.1×10^3 nm^2 considering the realistic tip-surface model and a lift height of 50 nm. This effective surface corresponds to a disc area with 95 nm in diameter which is inside the Ge island.

In fig.1, the diameter of these isolated Ge island is more than 100 nm. However, the SCM-PIT tip has a radius from 20 to 25 nm which is quite smaller than the island surface and the effective area is inside it. So these semi-sphere islands could be simplified as rectangular nano structures which has a flat surface interacting with the tip, see fig.4. This simple model could be used for quantifying trapped charge in oxide layers or embedded conductive nanostructures[15, 16].

Fig.4 Simplified nano-scale Ge island model

Using the parallel plate capacitor model, the total charge q inside the trapped area can be deduced as below:

$$q = \sqrt{\frac{\Delta\phi \times k \left(z_0 + \dfrac{d_{SiO_2}}{\varepsilon_{SiO_2}} + \dfrac{d_{Ge}}{\varepsilon_{Ge}} \right)^3 \varepsilon_0 S}{Q \left(\dfrac{d_{SiO_2}}{\varepsilon_{SiO_2}} + \dfrac{d_{Ge}}{\varepsilon_{Ge}} \right)^2}} \qquad (4)$$

$$D = \frac{q}{S} \qquad (5)$$

Where k is the cantilever spring constant, Q is the quality factor of the system and S is the effective tip area. D is the the charge density in this charged area.

In our experiment, the k is 1~5 N/m, Q is about 250, the dielectric constant for SiO_2 and Ge is separately 3.9 and 16.2, the charged Ge dot height is about 100 nm and S is 7.1×10^3 nm^2. Therefore, when the isolated Ge island is charged by -7V during 30 seconds, the injected charges density in the centre of the island is about 7.2×10^{-18} C and that at the edge is about 1.1×10^{-17} C. In the whole charge area, according to our calculation, there are an amount of 800 electrons were injected to this isolated nano-scale Ge island.

CONCLUSIONS

In summary, an individual Ge island on top of silicon dioxide layer has been charged by a conductive EFM tip and its charging behaviour has been quantitatively characterized at room temperature. Electrons or holes were successfully injected and were trapped homogenously in the isolated nano-scale Ge island. In order to quantitatively study these trapped charges, a truncated capacitor model was used to approximate the real capacitance between the tip and island surface. The analytical expression of the quantity of trapped charges in isolated Ge island as a function of the EFM phase signal was deduced. Applying a tip bias for -7V during 30 seconds leads to an injection about 800 electrons inside an individual Ge island. This characterisation work is of prime importance in developing electronics devices as memory transistors using Ge nano islands.

ACKNOWLEDGMENTS

Thanks X.Y. Ma for her helpful suggestions and Armel Descamps-Mandine from the CLYM platform facilities for his help and fruitful discussions on AFM measurements.

REFERENCES

1. G. Binning, C. Gerber and C. F. Quate, Phys. Rev. Lett. **56**, 930 (1989)
2. Dimension V Scanning Probe Microscope Manual from Vecco Company, 279 (2007)
3. R.M. Nyffenegger and R.M. Penner *Chem. Rev.* **97**, 1195 (1997)
4. D.M. Kolb, R. Ullmann and T. Will *Science* **275**, 1097 (1997)
5. J.W. Schultze and M.M. Lohrengel *Electrochim. Acta* **45**, 2499 (2000)
6. R. Berndt, R. Gaisch, J. K. Gimzewski, B. Reihl, R. R. Schlittler, W. D. Schneider, M. Tschudy, Science **262**, 1425 (1993).
7. P. Girard, Ph. Cadet, M. Ramonda, N. Shmidt, A. N. Usikov, W. V. Lundin, M. S. Dunaevskii, A. N. Titkov, Phys. Stat. Sol. (a), **195**, No.3, 508 (2003)
8. M. M. J. Bischoff, M. C. M. M. van der Wielen, H. van Kempen, Surf. Sci. **400**, 127 (1998).
9. K. Ito S. Ohyama, Y. Uehara, S. Ushioda, Surf. Sci. **324**, 282 (1995).
10. R. Dianoux, F. Martins, F. Marchi, C. Alandi, F. Comin, and J. Chevrier, Phys. Rev. B **68**, 045403 (2003)
11. Rachel A Oliver, *Rep. Prog. Phys.* **71**, 076501 (2008)
12. S. Sarid, Scanning Force Microscopy, Oxford University Press, New York, 1994.
13. S.Hudlet, M.Saint Jean, C.Guthmann, J.Berger, Eur. Phys. J.B **2**, 5 (1998)
14. F. Marchi, R. Dianoux, H.J.H. Smilde, P. Murd, F. Comin, J. Chevrier, J.Electrostatics **66**, 538 (2008).
15. D.Schaadt, E.T.Yu, S.Sanar, A.E.Berkowitz, Appl.Phys.Lett. **74**, 472 (1999)
16. C.Y. Ng, T.P. Chen, H.W. Lau, Y. Liu, M.S. Tse, O.K. Tan, V.S.W. Lim, Appl. Phys. Lett. **85**, 2941 (2004).

Magnetoresistive RAM (MRAM)

Mater. Res. Soc. Symp. Proc. Vol. 1250 © 2010 Materials Research Society   1250-G03-01

Fabrication of Prototype Magnetic Coupled Spin-Torque Devices for Non-volatile Logic Applications

Larkhoon Leem[1], James S. Harris[1], Charles Rettner[2], Brian Hughes[2], Xin Jiang[2], See-hun Yang[2] and Stuart Parkin[2]
[1]Department of Electrical Engineering, Stanford University, Stanford, CA 94305, U.S.A.
[2]IBM Almaden Research Center, San Jose, CA 95120, U.S.A.

ABSTRACT

A Magnetic Coupled Spin-torque Device (MCSTD) is a collective system of three interacting magnetic tunnel junctions (Mtjs) that forms a novel magnetic logic gate. The fundamental principle of the MCSTD is the modification of the energy barrier for spin-torque magnetization switching of a central (output) MTJ device arising from changes in the magnetic state of two input MTJ devices. The input MTJs are placed in close proximity of a few tens of nm of the output MTJ such that their magnetic fringing fields are strong enough (> 10 Oersted) to modulate the switching characteristics of the output device. By changing the magnetic states of the two input MTJs, four possible net magnetic fields at the center MTJ can be generated. A single MCSTD thereby enables NAND, NOR and XOR operations. In this paper, the fabrication of a prototype MCSTD device is described and preliminary experiment results are reported.

INTRODUCTION

Despite the discovery of a number of fundamentally new spintronic phenomena and major progress in our understanding of the basic physics, we are still far from demonstrating useful logic devices, which take advantage of spintronics. For example, for field effect transistors, which utilize spin currents, there are significant challenges in spin-injection, spin transport and spin detection and, moreover, in devising schemes that could enable them to compete in performance with conventional charge based devices. For these reasons, several "non-transistor" based spintronic logic devices have been proposed [1-7]. We have presented a new spintronics device architecture using Magnetic Coupled Spin-Torque Devices (MCSTD) [8] that falls into the second category in device architecture but operates by modulating the energy barrier needed to change the state of a device as in the first category. The MCSTD logic device has power gain and fan-out, and can implement the entire Boolean logic family of devices. In this paper, we summarize the mechanism of operation of the device, discuss the fabrication of a first prototype device, and present preliminary measurement data.

THEORY

Device operation mechanism

A Magnetic Coupled Spin-Torque Device (MCSTD) (see Fig. 1) consists of three magnetic tunnel junctions (Mtjs) that interact via magnetostatic fringing fields. A central MTJ is the *output* of the gate and two Mtjs on either side of it act as *inputs*. The main idea is that depending on the input MTJ magnetizations the magnetic reversal energy barrier of the output MTJ is modulated. Consequently, the current (or voltage) at which the central MTJ's magnetic state can be switched via a spin torque transfer mechanism is altered, and thereby the MCSTD can be used for logic operations.

A single MTJ with an elliptical cross section has a magnetic shape anisotropy that tends to align its magnetization along the major axis [9]. As a result, there are two stable states where

89

the magnetization is either pointing 'up' or 'down'. These two states can be mapped to logical '0' and '1' for logic operations. There is an energy barrier separating the two states, which has to be overcome to switch the magnetization. One possible way of building a magnetic logic device is by controlling this energy barrier with

Figure 1. Schematics of Magnetic Coupled Spin-Torque Device (MCSTD)

input signals. Components that set this barrier are (1) magnetic anisotropy (shape or crystalline) and (2) external magnetic field. While both of them can be used to control the energy barrier, we focus on utilizing magnetic fields from the adjacent MTJ devices to modulate the energy barrier of the switching device.

Logic devices that switch their magnetizations with current-induced magnetic fields have been proposed previously [7]. However, these devices do not scale well as has been realized for Stoner-Wohlfarth switched magnetic random access memories (MRAMs): the current and hence power to generate the magnetic fields that can address a single bit increases as the MTJ devices are scaled to smaller dimensions [10]. Our solution to generate magnetic fields that is compatible with device scaling comes from using additional spin-torque devices. If two additional devices are placed in the proximity of a few tens of nm to the original device (Fig. 2), their fringing magnetic fields are sufficiently strong to couple to the device and control the magnetic switching. This is a low-power solution as well because, the fringing fields are "free", i.e., no power is consumed to generate them, which is different from devices dependent upon current induced magnetic fields. Fig. 2 shows simulations [11] of the strength of the fringing fields when the input devices are aligned vs. anti-aligned. The free-layer of the MTJ is formed from a 2-nm thick CoFe layer. Magnetic fringing fields are maximum at the edges of the input device and quickly decay away from these edges. Nevertheless, the fringing fields are sufficiently large to affect the switching of neighboring MTJs if placed within a few tens of nms: the net fringing field at the center of the output device ranges from 1000 to 2800 A/m or from 13 to 36.4 Oe, when the distance between the two input devices is varied from 125 to 85nm.

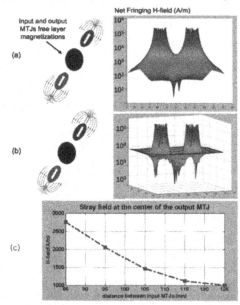

Figure 2. MCSTD NAND gate and its net magnetic stray fields from the input devices. (a) input device magnetizations in the same direction (b) in the opposite direction (c) Stray fields at the center of the output MTJ

90

Considering that the coercivity of spin-torque devices ranges from a few tens to hundreds of Oe, the fringing fields of this magnitude can clearly affect the switching behavior of the output MTJ device.
Different input MTJ magnetizations generate different net stray fields at the output MTJ (Fig. 2). If both input MTJs are magnetized in the same direction, their fields sum together to give a significant magnetic field at the center MTJ (Fig. 2(a)). If the input MTJs are magnetized in opposite directions, these fields cancel each other and have a reduced effect (Fig. 2(b)). The input MTJs in the MCSTD thus make the output MTJ easier or harder to switch when current is passed through the output MTJ: the biasing magnetic fields from the input MTJs work as a "third terminal", together with the two terminals which are used to pass current through the output MTJ, that controls the device operation. With strong magnetic coupling between the elements, the energy barrier height of the center MCSTD element can be modulated by the fringing fields of the input devices. This energy barrier modulation capability enables us to vary the switching voltage (ON vs. OFF) of a spin-torque device similar to that of the threshold voltage of a MOSFET. As a result, the MCSTD gate has an input signal dependent switching voltage. By judiciously choosing a voltage range that allows device switching only for specific inputs, we can build a generic logic family with only MCSTD gates. Details of functional logic gate designs such as NAND, NOR, NOT and XOR can be found in [8]. Entire Boolean logic functions are possible in MCSTD by changing, for example, the input MTJ location, angle and the output MTJ aspect ratio, as shown in Fig. 3.

Fabrication process

There are several challenges to the fabrication of MCSTD gates. These include: (1) patterning small area MTJs which are sufficiently close together (<30 nm) for strong magnetic coupling, but remain electrically isolated, (2) ion-milling the closely spaced MTJs whilst maintaining sidewalls of high integrity, (3) making electrical contacts to each of the closely spaced MTJs. In this section, we describe the fabrication process of MCSTD gates and our efforts to overcome these fabrication challenges.

First, the MTJ film stack is deposited by magnetron and ion beam sputtering. The fixed layer of the MTJ consists of an IrMn exchange bias layer and a CoFeB|CoFe|Ru|CoFe Synthetic Antiferromagnetic layer (SAF) [12]. Next, a mesa structure is patterned. For bottom contact, a group of nano-pillars (125 square pillars each sized $1um^2$) are patterned and etched with ion-milling technique. Since the voltage drop across this group of nano-pillars is negligible, the voltage level at the bottom contact can be read at the top of the nano-pillars.

Generally, negative e-beam resists such as Hydrogen silsesquioxane (HSQ) are used for nano-pillar e-beam lithography. However, HSQ was avoided in this work because of potential undesirable shrinkage. All of the following conditions can contribute to HSQ resist shrinkage: (1)

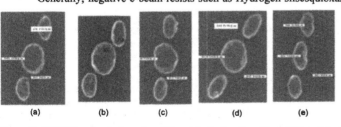

(a)   (b)   (c)   (d)   (e)

Figure 3. MCSTD gate designs for different logic functionalities. Different logic functions are derived from the difference in dipole couplings between the input and output device. Different input MTJ locations, angles and the output MTJ aspect ratios are tested.

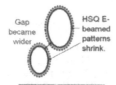

Gap became wider — HSQ E-beamed patterns shrink.

Chrome + Lift-off layers

Figure 4. Fabrication issue in using negative e-beam resist for MCSTD fabrication (top) Chrome hard mask layer as a solution (bottom)

time delay and ambient conditions between the e-beam exposure and subsequent development, (2) chemical cross-linking between HSQ and the underlying resist layer, (3) development time if the insoluble layer is removed by hydrofluoric (HF) acid, and, (4) long exposure time [13, 14]. Shrinkage of the HSQ resist is often favored to produce smaller dimension than what can be defined with e-beam lithography. However, HSQ shrinkage could lead to loss of control on the spacing between the MTJs (Fig.4). The spacing between the MTJ devices is a critical dimension of the MCSTD, which can affect the dipole coupling strength between these MTJs. In addition, fine details of the device patterns can be lost due to HSQ shrinkage. These effects will likely be more important when the MCSTD devices are scaled to smaller dimensions. To produce the effect of negative resist without using HSQ, positive resist such as PMMA was used as e-beam resist and Cr hard mask layer was prepared on top of each MTJs (Fig. 5). Appropriate thicknesses for the Cr mask layer and lift-off layer were found by trial-and-error: if the thickness of Cr mask layer is too small (e.g., <7nm), alumina sidewall will become too thick not to allow etchant to reach lift-off layer. Lift-off resist layer thickness was limited not to affect the e-beam lithography resolution.

A two-step ion-milling process was utilized to etch away the narrow gap between the input and output MTJs. Ion-beam angle iterated between the angle nearly perpendicular to the surface (for etching) and almost in parallel to the MTJ film surface (for removal of re-deposited material). Milling continued until IrMn layer was detected by in-situ Secondary Ion Mass Spectrometry (SIMS) analysis. The measured gap sizes between the MTJs were smaller than the nominal design spacing. This was due to the finite size of the electron beam and redeposition during the mask layer preparation process. Both effects increase the mask layer size and produce smaller spacing between the MTJs. The CAD design for MCSTD e-beam lithography should take this effect into account and start from a larger gap size design to achieve the intended gap spacing. Finally, electrical contact lines were placed on the top of each MTJ device. Due to the proximity of the contact lines, this process was carried out by a combination of e-beam and optical lithography (Fig.6).

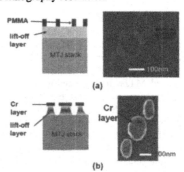

Figure 5. E-beam lithography and Cr hard mask layer preparation for ion-milling

DISCUSSION

Functional MCSTD devices with 20 nm gaps between the input and output MTJ devices were successfully fabricated. These devices demonstrated the input signal dependence of the switching voltage. Fig. 7 shows typical magneto-resistance versus voltage (R-V) loops of three MCSTD gates. The loops show clear hysteresis. In this experiment, logical '1' is defined as magnetization upward and '0' with magnetization downwards. In the upper scans, the input devices have (1,1) input and in the lower they have (0,0) input. The devices both show the shifts

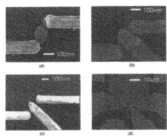

Figure 6. SEM images of electrical contacts on the output and input MTJs of MCSTD gates

in the switching voltage and the expected magnetic reversal when the input signals, i.e., the input MTJ magnetizations change. The R-V loops from 30 iterations of MCSTD gate switching were accumulated to confirm the reproducibility of switching voltage shifts. In all cases, the voltage shifts were larger than thermal fluctuations: in Fig. 7(d), we measured the standard deviation of the switching voltage shifts of the MCSTD gate shown in Fig. 7(c) with respect to the scanned background magnetic field. This was to show the strength of magnetic coupling between the input and output devices and the thermal fluctuations in the switching voltage shifts. If the magnetic coupling were weak, the standard deviation of the voltage shift would be affected by the large background magnetic fields and change. However, the experimental results showed no dependency on background magnetic fields, which confirms strong coupling between devices. The average value of the thermal fluctuations in the switching voltage, 26.5mV, was much

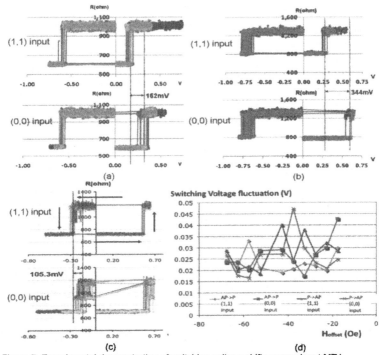

Figure 7. Experimental demonstration of switching voltage shifts versus input MTJ magnetizations

93

smaller than the magnitude of the voltage shifts. The MTJ devices in Fig.7 have the following properties: H_c=44~80 Oe, TMR=46~78.4%, RA=2.1~8.3 ohm.cm^2. The input and output MTJ sizes are 53×110 nm^2 and 100×140nm^2.

Since the devices interact via their fringing fields, it is very important to maintain the uniformity in device shape and distance between the input and output MTJs. However, initial experimental results suggest that such processing limitations are well within the current technological limits of our processing. Furthermore, adopting other means such as perpendicular magnetic anisotropy instead of shape anisotropy to magnetically couple the devices can mitigate the device sensitivity to size and shape.

CONCLUSIONS

Magnetic Coupled Spin-torque Device (MCSTD) enables a novel spintronics device architecture that uses spin current controlled magnetic couplings to modulate the energy barrier of an MTJ gate. By taking advantage of magnetic fringing fields, which are innately low energy interactions, MCSTD devices consume very low energies if the voltages needed to switch and operate the MTJ elements are made very small. We have demonstrated the feasibility of fabricating MCSTD prototype devices and have also shown experimentally that they operate as anticipated by micromagnetic simulations.

REFERENCES

[1] Allwood, D. A., et al. Magnetic domain-wall logic. Science 309, 1688–1692 (2002)
[2] Behin-Aein, B., Datta, D., Salahuddin, S., Datta, S. Proposal for an all-spin logic device with built-in memory, *Nature Nanotechnology*, DOI: 10.1038 (2010)
[3] Carlton, D. B., Emley, N. C., Tuchfeld, E. Bokor, J. Simulation of nanomagnetbased logic architecture. *Nano Lett.* **8**, 4173–4178 (2008)
[4] Cowburn, R. P., Welland, M. E. Room temperature magnetic quantum cellular automata. *Science* **287**, 1466–1468 (2000)
[5] Imre, A., et al. Majority logic gate for magnetic quantum-dot cellular automata. *Science* **311**, 205–208 (2006)
[6] Khitun, A., et al. Spin wave logic circuit on silicon platform. *Fifth International Conference on Information Technology: New Generations,* 1107–1110 (2008)
[7] Ney, A., et al. Programmable computing with a single magnetoresistive element. *Nature* **425**, 485–487 (2003)
[8] Leem, L., Harris, J. S. Magnetic coupled spin-torque devices and magnetic ring oscillator. *Proc. IEDM* doi: 10.1109/IEDM.2008.4796640 (2008)
[9] Leem, L., Harris, J. S. Magnetic coupled spin-torque devices for nonvolatile logic applications. *J.Appl.Phys.* **105**, 07D102 (2009)
[10] Chen, E., Current Status and Future Outlook of STT-RAM Technology. *18th Information Storage Industry Consortium Annual Meeting* (2008)
[11] Donahue, M. J. and Porter, D. G. and National Institute of Standards and Technology (U.S.) OOMMF user's guide (1999)
[12] Parkin, S.S.P. Spin-Polarized Current in Spin Valves and Magnetic Tunnel Junctions. *MRS Bulletin* **31**, (2006)
[13] Driskill-Smith, A.A.G., et al. Electron-beam lithography for the magnetic recording industry: Fabrication of nanoscale (10nm) thin-film read heads. *Microelectronic Engineering*, **73-74** 547-552 (2004)
[14] Personal communication with Charles Rettner (2010)

Mater. Res. Soc. Symp. Proc. Vol. 1250 © 2010 Materials Research Society
1250-G03-02

Structural, Magnetic and Magneto-transport Properties of Reactive-sputtered Fe_3O_4 Thin Films

X.H. Wang[1], S. Goolaup[1]., P. Ren[1], and W.S. Lew[1],
[1]School of Physical and Mathematical Sciences, Nanyang Technological University, 637371, Singapore

ABSTRACT

Thin films of magnetite (Fe_3O_4) are grown on a single-crystal Si/SiO_2 (100) substrate with native oxide using DC reactive sputtering technique at room tempreture (RT) and 300°C. The x-ray diffraction(XRD) result shows the thermal energy during deposition enhances the crystallization of the Fe_3O_4 and x-ray photoelectron spectroscopy confirms the film deposited at 300°C is single-phase Fe_3O_4 while the film deposited at RT is mostly γ-Fe_2O_3. The electrical measurements show that the resistivity of the Fe_3O_4 film increases exponentially with decreasing temperature, and exhibits a sharp metal-insulator transition at around 100 K, indicating the Verwey transition feature. The saturation magnetization M_s of Fe_3O_4 film measured by vibrating sample measurement (VSM) at RT was found to be 445 emu/cm^3.

INTRODUCTION

The field of spintronics [1] has attracted great research effort recently, due to the potential applications in various spin-based devices [1]. The injection and detection of spin current which can combine with the magnetic recording technique draws large amount of attention to high spin polarized ferromagnetic material[2]. Half metallic materials whose charge carriers are fully polarized at Fermi level have been proposed candidates in magnetoresistive devices such as giant magnetoresistance (GMR) multilayers, spin valves and magnetic tunnel junctions. Among the possible half metallic compound such as Fe_3O_4, $La_{0.7}$ $Sr_{0.3}$ MnO_3 and CrO_2, Fe_3O_4 is candidate for future spintronics devices due to the high Curie temperature of 858K[3] and exhibition of full negative spin polarization[4] at room temperature. In order to obtain optimal performance in the applications, Fe3O4 thin film growth with high quality is required. Different techniques have been used such as: molecular beam epitaxy[5], pulsed laser deposition, and sputtering[6]. Various substrates such as MgO, $MgAl_2O_4$, $SrTiO_3$, sapphire and Si have been used to study the influence of the lattice mismatch between the magnetite and the substrates.

Even though, Fe_3O_4 is very promising for spintronic devices, it has not been so far optimized for application in this field due to its poor inclusions in device architectures mostly technologically fabricated on semiconducting substrates like Si or GaAs. Fe_3O_4 films deposited directly on Si substrates, Si with buffer layers as well as GaAs using electron beam evaporation[7] and laser ablation[8] have been studied. In this work, considering the compatibility with most of the devices, magnetite thin films were deposited on surface of SiO_2/Si

using DC reactive magnetron sputtering. The Verwey transition[9] feature indicated in the electrical measurements shows that high quality Fe_3O_4 thin films have been produced.

EXPERIMENTAL DETAILS

Fe_3O_4 films were grown by DC magnetron sputtering on thermally oxidized Si(100) substrate at a process pressure of 3×10^{-3} Torr. Prior to deposition, the Si(100) substrates were first rinsed in acetone, 2-isopronal and de-ionized water. A buffer layer comprising of 5 nm thick Ta layer was then grown on the substrate. The Fe_3O_4 films were reactively sputtered from pure Fe (99.99%) target in an Ar-O_2 (27:6.8) mixture, with the substrate maintained at room temperature (sample A) and 300C (sample B), respectively. The base pressure was better than 5×10^{-7} Torr. The film thickness grown was kept at 60nm for both set of samples. Structural characterization of the deposited samples was carried out using conventional θ-2θ X-ray diffraction (XRD) and X-ray photoelectron spectroscopy (XPS). The magnetic and transport measurements were conducted using vibrating sample magnetometer (VSM). For electrical measurements, contacts were made using standard optical lithography, metallization and lift off of 10 nm Cr/ 150 nm Au. The resistance was recorded using a two terminal method. Low temperature measurements were performed using a physical measurement system (PPMS-Quantum Design).

RESULTS AND DISCUSSIONS

Shown in Fig 1(a), is the typical θ-2θ XRD pattern for sample A, Fe_3O_4 film gown onto substrate maintained at RT. The XRD pattern displays a single distinct Fe peaks in the (440) orientation, indicating that the polycrystalline Fe_3O_4 film has a preferred growth along the 110 orientation. We attribute this preferred orientation to the columnar growth structure of the film due the sputtering deposition technique [10]. The XRD pattern for sample B, on the other hand, displays four Fe peaks along the (311), (511), (220) and (400) orientations, as seen in Fig 1(b). Interestingly, the (440) peak, as seen in sample A, is totally suppressed. The additional peaks, as seen from the XRD patterns, indicate a structural change in the Fe_3O_4 film grown at 300C. This may be due to the thermal energy from the substrate, promoting different film growth mechanisms.

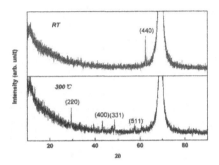

Fig 1. XRD patterns of 60 nm thick Fe_3O_4 films on Si/SiO_2 (100) wafers at RT and 300°C.

The reactive sputtering of Fe with Ar and O_2 plasma, may result in the formation of either or both of the two stable Fe oxides, namely γ-Fe_2O_3 and Fe_3O_4. These two Fe oxides, share the same spinel structure, and as such cannot be distinguished in XRD measurement. To differentiate between the Fe_3O_4 and γ-Fe_2O_3 that may have formed during the reactive sputtering process, we conducted XPS scan on Fe2p using MgK radiation at 1256.3 eV. The Fe_{2p} spectrum of γ-Fe_2O_3 exhibits a satellite peak at 719eV, while the Fe_{2p} spectrum of Fe_3O_4 is demarcated by a shoulder at 709 eV. The XPS scans for both samples, A and B, are shown in Figure 2. From Fig 2(a), we observed that the XPS of the sample A displays both a shoulder at 709eV and a peak at 717eV. This implies that the film grown on sample A may comprises both the formation of Fe_2O_3 and Fe_3O_4 state during growth. For Sample B, the peak at 717 eV is absent, as seen in Fig 2(b), implying that the film grown on sample A consists of only one oxide of Fe, Fe_3O_4. The shift of the peak from 719 eV may possibly due to the formation of other phase of iron oxide in sample

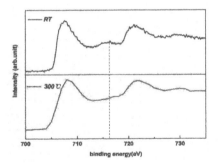

97

Fig. 2. Fe XPS spectra for samples deposited on Si/SiO$_2$ at RT and 300°C

To better understand the magnetic properties of the samples, we carried out magnetization loop measurement using VSM. The room temperature magnetization loops of the two sets of samples, measured with magnetic field applied parallel to the film plane, are presented in Fig 3. The inset in Fig 3 show the normalized M-H loops for both sets of samples. From Figure 3, we can see that sample B exhibits a much larger saturation magnetization of 445 emu/cm^3, as compared to sample A, 134 emu/cm^3. Fe$_2$O$_3$ is known to be diamagnetic and the contribution to the overall magnetization is negligible. As such, the total magnetic moment for both samples A and B are solely from the Fe$_3$O$_4$ in the film. The drastic increase in the saturation magnetization, for sample B, is due to the suppression of Fe$_2$O$_3$ phase formation, when the substrate is maintained at 300C. Interestingly, the saturation magnetization for sample B, is close to the value of a single crystal Fe$_3$O$_4$, which is reported to be 471 emu/cm^3 [11]. We attribute the high saturation magnetization of our film to the good structural quality. Another reason for the enhancement of the magnetization of Fe$_3$O$_4$ is possibly due to the thermal energy contributing to the crystallization of the Fe$_3$O$_4$. This in turn reduces the thickness of amorphous Fe$_3$O$_4$ layer which is formed at the surface of the substrate due to the lattice mismatch between the substrate and the Fe$_3$O$_4$. As seen from the inset in Fig 3, sample A exhibits a much larger saturation field of 2000 Oe as compared to that of sample B, which is only 500 Oe. The high saturation field of the film A is possibly due to the field irreversibility of the Fe$_2$O$_3$.

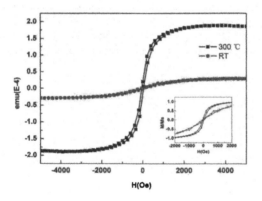

Fig. 3. Hysteresis loops measured at 300K for samples deposited at RT and 300°C

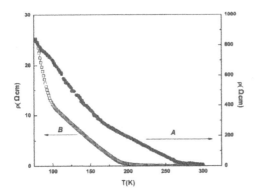

Fig. 4. Electrical resistivity of a 60-nm-thick sputtered Fe_3O_4 film grown on Si/SiO_2 substrates plotted vs T. The Verwey transition can be clear seen in the film deposited at 300 °C but absent in the film deposited at RT.

We have further carried out transport measurement, on the samples. Shown in Fig.4, is the resistivity of both sets of samples as function of temperature. For both sample A and B, there is an exponential increase of resistivity as the temperature is reduced. Sample A display a room temperature resistivity of ~120K. When the temperature is reduced below 270 K, there is a gradual increase in resistance with decreasing temperature. For sample B, room temperature resistivity is 0.005292 Ω cm. As the temperature is reduced below 200K, we observed a gradual increase in resistivity with temperature. At 100K, a sharp increase in resistivity is observed, indicating a metal-insulator transition, a clear signature of the Verwey transition feature. The temperature discrepancy of the verwey transition from that of bulk Fe3O4, ,is due to cation vacancy[12] and the nonstoichiometric[13] deviation.

While for the film A, the resistivity increases significantly and the Verwey transition is absent. It is possibly due to the formation of γ-Fe_2O_3 which makes resistivity decrease and the Verwey transition smeared out in the curve as compared to B. The Verwey transition and relatively high magnetization in the film B show we have successfully fabricated the Fe_3O_4 at RT on Si substrate using sputtering technique.

CONCLUSIONS

In summary, we have grown Fe_3O_4 thin film on thermally oxidized Si substrate maintained at different temperatures. The film was grown using a Fe target in an ArO_2 plasma, using DC reactive sputtering. We observed that the sample with substrate maintained at 300C, has a very large magnetic moment of 445 emu/ cm^3. This was attributed to the suppression of the Fe_2O_3

phase, due to the substrate temperature promoting the crystallization of Fe_3O_4. Our transport measurement reveals a clear signature of verwey transition at 100K, for sample grown at 300C. The presence of the verwey transition indicates that our Fe_3O_4 thin film is our very good structural quality.

REFERENCES

[1]   Y. Ohno, D. K. Young, B. Beschoten, F. Matsukura, H. Ohno, and D. D. Awschalom, *Nature*, 402, 790, (1999).

[2]   M. Okada, S. Nagai, Y. Neo, K. Hata, and H. Mimura in *21st International Microprocesses and Nanotechnology Conference*, Fukuoka, JAPAN, (2008).

[3]   A. V. Ramos, S. Matzen, J. B. Moussy, F. Ott, and M. Viret, *Phy. Rev. B*, 79, (2009).

[4]   M. Bohra, S. Prasad, N. Venketaramani, N. Kumar, S. C. Sahoo, and R. Krishnan, *J. Mag. Magn, Mater.*, 321, pp. 3738-3741,(2009).

[5]   C. Magen, E. Snoeck, U. Luders, and J. F. Bobo, *J. Appl. Phys.*, 104, (2008).

[6]   M. Jung, S. Park, D. Park, and S. R. Lee, *J. Korean Institute of Metals and Materials*, vol. 47, pp. 378-382, Jun 2009.

[7]   S. Jain, A. O. Adeyeye, and C. B. Boothroyd, *J. Appl. Phys.*, 97, May (2005).

[8]   R. J. Kennedy and P. A. Stampe, *J. of Phys. D-Appl. Phys.*,. 32, pp. 16-21,(1999).

[9]   Z. Zhang and S. Satpathy, *Phys. Rev. B*, 44, p. 13319, (1991).

[10]   Y. G. Peng, C. Park, and D. E. Laughlin, *J. Appl. Phys.*, 93, pp. 7957-7959, (2003).

[11]   D. Reisinger, P. Majewski, M. Opel, L. Alff, and R. Gross, *Appl. Phys. Lett.*, 85, pp. 4980-4982, (2004).

[12]   J. B. Yang, J. Zhou, W. B. Yelon, W. J. James, Q. Cai, K. V. Gopalarkishnan, S. K. Malik, X. C. Sun, and D. E. Nikles, *J. Appl. Phys.*, 95, pp. 7540-7542, (2004).

[13]   J. P. Shepherd, J. W. Koenitzer, R. Aragon, J. Spal/ek, and J. M. Honig, *Phys. Rev. B*, 43, p. 8461, (1991).

Mater. Res. Soc. Symp. Proc. Vol. 1250 © 2010 Materials Research Society   1250-G08-02

Low Temperature Deposition of Ferromagnetic Ni-Mn-Ga Thin Films From Two Different Targets via rf Magnetron Sputtering

A. C. Lourenço[1], F. Figueiras[1], S. Das[1], J. S. Amaral[1,5], G. N. Kakazei[5], D. V. Karpinsky[1], N. Soares[1], M. Peres[2], M. J. Pereira[1], N. M. Santos[2], P. B. Tavares[4], N. A. Sobolev[2], V. S. Amaral[1] and A. L. Kholkin[3]

[1]Department of Physics, CICECO, University of Aveiro, 3810-193 Aveiro, Portugal

[2]Department of Physics, I3N, University of Aveiro, 3810-193 Aveiro, Portugal

[3]Department of Ceramics and Glass Engineering, CICECO, University of Aveiro, 3810-193 Aveiro, Portugal

[4]Centro de Química, Universidade de Trás-os-Montes e Alto Douro, 5001-801 Vila Real, Portugal

[5]IFIMUP-IN and Departamento de Física e Astronomia da Faculdade de Ciências da Universidade do Porto, Rua do Campo Alegre 687, 4169-007 Porto, Portugal

ABSTRACT

Low temperature (400°C) deposition of ferromagnetic Ni-Mn-Ga thin films is successfully performed via rf magnetron sputtering technique using co-deposition of two targets $Ni_{50}Mn_{50}$ and $Ni_{50}Ga_{50}$ on sapphire (0001) and Si (100) substrates. The films are in part amorphous with significant degree of crystallinity. The obtained crystallographic structure is shown to be substrate-dependent. Films on both substrates are ferromagnetic at room temperature (Curie temperature ~ 332 K) demonstrating well-defined hysteresis loops, low coercivity (~ 100 Oe), and saturation magnetization of ~ 200 emu/cc. At low temperature (5 K), both films are characterized by increased magnetization and wider hysteresis loops with higher coercivity and remanent magnetization. The process is therefore effective in achieving the appropriate thermodynamic conditions to deposit thin films of the Ni-Mn-Ga austenitic phase (highly magnetic at room temperature) at relatively low substrate temperature without the need for post-deposition annealing or further thermal treatment, which is prerequisite for the device fabrication.

INTRODUCTION

Ni–Mn–Ga thin films hold great promise in obtaining a shape memory effect with a high power density and a low driving field. The films can be produced by different techniques like sputtering [1], pulsed laser deposition [2-3], molecular beam epitaxy [4], or flash evaporation [5]. However, a magnetically induced reorientation of the martensite variants is observed only in few cases. Recent work by Heczko et al. [6] showed a distinct room temperature shape memory effect in a substrate-constrained film, which was attributed to the magnetic field induced reorientation of a small percentage of the twin variants. These results provide us the evidence and requirement for developing the growth/preparation conditions to optimize the shape memory effect by improving the crystallinity and microstructure of these films.

Ni–Mn–Ga films having room temperature ferromagnetism needs sufficiently high deposition temperatures [2], which, on the other hand, disrupts the corresponding stoichiometric compositions. General approach is therefore taken by depositing the films at low temperature followed by high temperature annealing [7]. However the post-growth heat treatment has certain limitation regarding the maintaining of uniform stoichiometry throughout the film.

In this paper, we report the *in-situ* growth of ferromagnetic Ni-Mn-Ga films at low temperature (400 °C) in a custom designed magnetron rf-sputtering system using co-deposition of two targets $Ni_{50}Mn_{50}$ and $Ni_{50}Ga_{50}$ on sapphire (0001) and Si (100) substrates.

EXPERIMENT & DISCUSSION

Ni–Mn–The spherical sputtering chamber used here is 30 cm in diameter and hosts three 2" planar magnetron guns mounted in the co-focal horizontal on-axis geometry for sequential and/or co-deposition. In order to maintain the proper stoichiometric ratio, the gas pressure is controlled by the conventional rotary/turbo pump setup and measured with a combined Pirani-Hot Cathode gauges (base pressure at room temperature $\leq 5.0 \times 10^{-8}$ mbar for all films). Two independent MKS mass flow controllers (100 sccm) are used to control Ar composition during deposition, from 1.0×10^{-5} mbar up to 10^{-1} mbar. A commercial 2" diameter resistive heater is used to heat the substrates up to 950 °C and using special steel masks it is capable of accommodate 4 different $10 \times 10 \times 0.5$ mm^3 substrates, for each deposition run. The temperature stability is ±10 °C at 850 °C. The commercially available substrates are cleaned in ultrasonic baths with acetone, ethanol and deionized water before deposition. Two different commercial $Ni_{50}Mn_{50}$ and $Ni_{50}Ga_{50}$ bulk targets (99.9% purity) were simultaneously sputtered in order to grow thin films with composition close to stoichiometric Ni_2MnGa. The deposition rates and composition of the two components were studied by separately depositing the thin films of Ni-Mn and Ni-Ga on different substrates in order to study the mass effect [8] and deposition geometry on film composition [9].

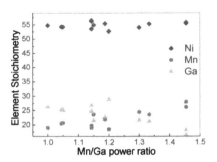

Figure 1. Variation of Mn/Ga power ratio on the Ni-Mn-Ga thin film using co-deposition process.

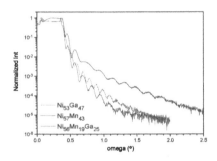

Figure 2. X-ray reflectometry of deposited thin films of $Ni_{53}Ga_{47}$, $Ni_{57}Mn_{43}$, and $Ni_{56}Mn_{19}Ga_{25}$.

In Figure 1 we can see the effect of the applied Mn/Ga power ratio on the stoichiometry of Ni-Mn-Ga thin films. It is clear from the Figure that the co-focal horizontal on-axis geometry is able to effectively control the Mn/Ga ratio by controlling the magnetron powers. Figure 2 shows the X-ray reflectivity curves for the individual components of Ni-Ga and Ni-Mn and for a Ni-Mn-Ga film developed under simultaneous deposition. The growth rate is ~ 2 nm/min. The thickness of the films obtained from X-ray reflectivity curves comes out to be ~ 120 nm. SEM cross-section of the films reveals the columnar growth of both films (on sapphire and Si). Energy dispersive X-ray spectroscopy confers the atomic composition of the films as approximately $Ni_{56}Mn_{19}Ga_{25}$. The room temperature X-ray diffraction shows (Figure 3) the co-existence of a distinct amorphous structure with some degree of crystallinity for both films under study. A clear evidence of both cubic and tetragonal phases is obtained for NMG film grown on sapphire, whereas mainly tetragonal phase (with small amount of cubic phase) is observed in films deposited on Si.

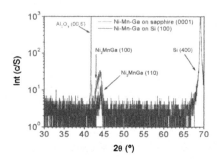

Figure 3. X-ray Bragg diffraction pattern of the deposited Ni-Mn-Ga films on sapphire and on Si.

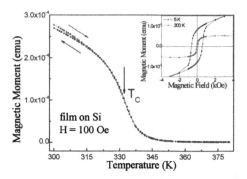

Figure 4. The magnetization data of a represenative Ni-Mn-Ga film on Si. The inset shows the magnetic hysteresis curves at 5 K and at room temperature.

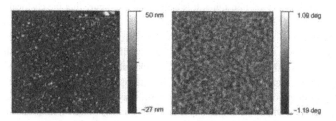

Figure 5. The topography (left) and MFM phase image of a representative Ni-Mn-Ga film.

It is interesting that the difference in crystal structure observed in films deposited on both substrates does not influence magnetic phase transition and our SQUID measurements indicate Curie temperatures to be about 332 K for the investigated Ni-Mn-

104

Ga films (Figure 4). The room temperature ferromagnetic behavior is characterized with well-defined hysteresis loops of similar shape. The coercivity value obtained is ~100 Oe with saturation magnetization of the order of 200 emu/cc. This corresponds to about 25 emu/g (assuming bulk density of 8.16 gm/cc for the films), a value about half of magnetization in bulk stoichiometric Ni-Mn-Ga in martensite phase [10]. We argue that lower magnetization value pertains to the observed amorphous structure of these films and, in its turn, is apparently a result of low temperature deposition. Magnetic Force Microscopy data (Figure 5) presents clear magnetic domain patterns common for the perpendicular magnetic anisotropy of the films. The observed domains configuration is characteristic of large magnetocrystalline anisotropy. At low temperature (5 K), the hysteresis loop gives higher values of both coercivity and remnant magnetization in Ni-Mn-Ga films (Figure 4). Detailed magnetization experiments are however necessary for the understanding of the structural phase transition in Ni-Mn-Ga films.

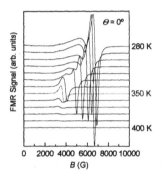 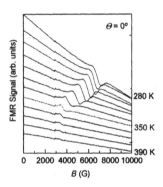

Figure 6. Ferromagnetic resonance spectra of the Ni-Mn-Ga film on sapphire (left) and that on Si (right) for magnetic field perpendicular to the sample plane ($\Theta=0°$).

In addition, we performed magnetic resonance (MR) measurements in the X-band (~9.47 GHz) at dc magnetic fields (B_0) in the 0-1.5 T range and in the temperature range from 280 K to 400 K. The MR curves were obtained by the magnetic field modulation technique with a modulation frequency of 100 kHz, so that the first derivative of the absorption spectrum was registered. A clear evolution from ferromagnetic resonance (FMR) to paramagnetic resonance (EPR) was observed with increasing temperature in both films (Figure 6), which is related to the observed ferromagnetic phase transition. The spectra reveal two magnetic phases in the film grown on sapphire while only one in that on Si. Moreover, the line intensity in the former sample is substantially higher than in the latter one. This is probably caused by the microwave absorption by free charge carriers in the Si substrate.

For both cases, the resonance field is a maximum for the field orientation perpendicular to the film plane. This means that it is the shape anisotropy that determines the (in-plane) orientation of the easy axis. No in-plane anisotropy in FMR spectra was detected. The magnetization of the magnetic phase giving rise to the FMR absorption can be measured in the first approximation as a field difference between the EPR line position for $g = 2$ ($B_0 \approx 0.33$ T) and the actual FMR line position. For the narrower line in the film grown on Al_2O_3 and for the unique line in the film grown on Si this difference is approximately the same, e.g., $B_0 \approx 0.68$ T at 280 K. The peak-to-peak width of the narrow line in the film grown on sapphire ($\Delta B_{pp} \approx 0.08$ T at $T = 280$ K) is about 2.5 times larger than that in the film grown on Si ($\Delta B_{pp} \approx 0.032$ T at the same temperature), which indicates a better uniformity of the film grown on sapphire.

CONCLUSIONS

In conclusion, we used co-focal horizontal on-axis geometry in our specially designed magnetron sputtering system in order to control the thermodynamic conditions for the deposition of ferromagnetic Ni-Mn-Ga thin films with proper stoichiometry at low temperatures. The process eliminates the need of any further post-deposition annealing or thermal treatment and is highly relevant for the device fabrication. The work is currently underway to deposit Ni-Mn-Ga films with better crystallinity and on "active" perovskite substrates for the use in memory applications.

ACKNOWLEDGEMENTS

The work was performed within the EC-funded project "Multiceral" (NMP3-CT-2006-032616). Partial financial support from FCT is also acknowledged (SFRH/BD/25011/2005, SFRH/BPD/63942/2009, SFRH/BPD/39262/2007, and SFRH/BPD/ 42506/2007).

REFERENCES

1. M. Thomas, O. Heczko, J. Buschbeck, U. K. Rossler, J. McCord, N. Scheerbaum, L. Schultz, and S. Fahler, *New J. Phys.* **10**, 023040 (2008).
2. A. Hakola, O. Heczko, A. Jaakkola, T. Kajava and K. Ullakko, *Appl. Phys.* A **79** 1505 (2004)
3. V. O. Golub, A. Y. Vovk, L. Malkinski, C. J. O'Connor, Z. Wang, and J. Tang, *J. Appl. Phys.* **96** 3865 (2004)
4. J. W. Dong, L. C. Chen, J. Q. Xie, T. A. R Muller, D. M. Carr, C. J. Palmstrom, S. McKernan, Q. Pan, and R. D. James, *J. Appl. Phys.* **88** 7357 (2000)
5. J. Dubowik, Y.V. Kudryavtsev and Y.P. Lee, *J. Magn. Magn. Mater.* **272** 1178 (2004)
6. O. Heczko, M. Thomas, J. Buschbeck, L. Schultz and S. Fahler, *Appl. Phys. Lett.* **92** 072502 (2008)
7. V.A. Chernenko, M. Ohtsuka, M. Kohl, V.V. Khovailo and T. Takagi, *Smart Mater. Struct.* **14** S245 (2005)
8. R.R. Olson, M.E. Kingand and G.K. Wehner, *J. Appl. Phys.* **50** 3677 (1979)
9. US Patent no. US 6454913 B1 (Sept 24, 2002)
10. A. Sozinov, A. A. Likhachev, and K. Ullakko, *IEEE Trans. Magn.* **38** 2814 (2002)

Mater. Res. Soc. Symp. Proc. Vol. 1250 © 2010 Materials Research Society   1250-G08-03

Magnetostatic Interactions of Two-Dimensional Arrays of Magnetic Strips.

Leszek M. Malkinski, Minghui Yu and Donald J. Scherer II

Advanced Materials Research Institute, University of New Orleans, New Orleans, LA 70148, U.S.A

ABSTRACT

A series of arrays consisting of Permalloy stripes with dimensions of 100 nm x 300 nm x 1500 nm were fabricated using electron beam nanolithography and magnetron sputtering followed by the lift-off process. In order to elucidate the effect of magnetostatic interactions among nanosized stripes on magnetic properties of the arrays, the separation between the stripes in different arrays was varied in the range between 100 nm and 2000 nm. Magnetic hysteresis loops of the arrays were measured using SQUID magnetometer for different orientations of the applied field with respect to the arrays. Magnetic anisotropy of the arrays was determined based on ferromagnetic resonance measurements at 9.8 GHz using EPR spectrometer. The measurements were carried out for different directions of the in-plane magnetic bias filed. The angular dependence of the resonance field of the main resonant peak indicated presence of the uniaxial magnetic anisotropy due to elongated shape of the stripes. Comparison between angular curves of resonant fields for different arrays leads to the conclusion that increasing strength of magnetostatic interactions among the stripes leads to a suppression of the uniaxial anisotropy. The stripes separated by 2000 nm behave almost like non-interacting objects, but the effect of interactions becomes particularly significant for separations smaller than 600 nm. The properties of the arrays with the smallest separations resembled those of continuous films. Magnetostatic modes have been observed in the FMR spectra in addition to the main resonant peak. These modes are believed to result from dimensional confinement of lateral spinwaves in the magnetic stripes. No such modes were observed in the reference samples of solid Py films, with the in-plane applied magnetic field.

INTRODUCTION

Dynamically developing applications, in the fields of wireless communication and information technology, require advanced magnetic materials operating at microwave frequencies [1-4]. Technology of conventional magnetic microwave materials with complex structures, such as: spinels, hexaferrites or garnets, cannot be easily integrated with current silicone technology of electronic circuits. Therefore, there is an increasing demand for developing new materials for microwave applications. Metallic nanostructures of transition metals with large magnetization seem to be good candidates for high frequency devices. The problem of losses, due to eddy currents, which limit frequency of applications of bulk metals, is greatly reduced because of reduced dimensionality of nanostructures. Our preliminary results on arrays of magnetic stripes or antidots indicate that magnetostatic fields have pronounced effect on their magnetic behavior at microwave region [5-7]. One of the effects of magnetostatic fields is the shape anisotropy resulting from different demagnetization factors for different orientations of the nanoparticles with respect to the applied field. Magnetic stripes with different aspect ratios of length to width displayed significant differences in their response to microwave fields. Another effect is magnetostatic interactions between fields produced by the neighboring nanosized objects. The aim of this work is to discuss the influence of interactions among nanosized magnetic stripes in arrays on their static and dynamic magnetic properties.

EXPERIMENTAL

Electron beam nanolithography in combination with sputtering and lift off technique was used to fabricate a series of arrays of stripes with the following dimensions: height h=100 nm, width w=300 nm and length l = 1500 nm from Permalloy with nominal composition of 81% Ni and 19% Fe. The patterned area of 1.65 mm x 1.65 mm was subdivided into an array of 15 x 15 identical patches with dimensions of about 100 μm x 100 μm each. Longer edges of five samples were separated by distances of 100, 300, 600, 1000 and 2000 nm, respectively. The separation between the shorter edges (rows of stripes) in these samples was kept constant at 1000 nm. The sixth sample was fabricated with all separations of 100 nm. Both field emission microscopy and atomic force microscopy were used to test the quality of the patterns. Examples of fragments of field emission scanning electron microscope images of the arrays are presented in Fig.1. SQUID susceptometer type MPMS XL from Quantum Design was used to measure room temperature magnetic hysteresis loops. Absorption of microwave radiation by the arrays was studied at room temperature by the means of ferromagnetic resonance technique (FMR) using an X-band Bruker EMX300 electron paramagnetic resonance (EPR) spectrometer.

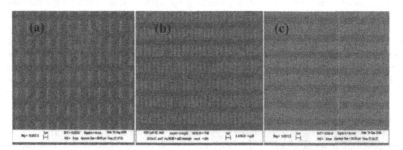

Fig. 1. Three examples of field emission scanning electron microscope images of arrays of Py stripes (100 nmx300 nmx1500nm) separated by distances of 2000 nm (a), 600 nm (b) and 100 nm (c) Rows of the particles are separated by 1000 nm in both samples.

RESULTS AND DISCUSSION

Because of elongated shape of nanosized stripes of magnetically soft Py, the magnetization process depends on the direction of the applied field with respect to the sample. The angular dependence of the magnetization processes is typical of uniaxial anisotropy with the easy magnetization direction along long axis of the stripes and the hard anisotropy axis transverse to it. Because this anisotropy is related to the shape of the objects, it is also known as "shape anisotropy". Our former studies of the arrays of stripes demonstrated that both magnetization hysteresis loops and ferromagnetic resonance fields strongly depend on the aspect ratio of the width to length of the stripes. In order to separate this effect from interactions we selected only one type of stripes with aspect ratio 1:5 and fixed dimensions (100x300x1500 nm). We focused on the effect of magnetostatic interactions among nanostripes in the arrays with different separations between the stripes.

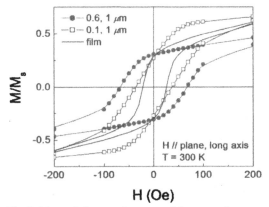

Fig. 2. Magnetic hysteresis curves of three samples: two arrays of stripes with separations of 600 nm (solid circle) and 100 nm (open square), and the reference sample of 100 nm thick Py film (solid line). The measurements were carried out with the field aligned with the long edges of the stripes.

Typical hysteresis curves measured along the easy magnetization direction for two arrays with stripes separated by 600 nm and 100 nm are compared in Fig. 2 with the loop for a continuous film. The coercivity of the patterned arrays was significantly larger than that of the film and it was decreasing with decreasing the separation between the stripes. The evolution of the ferromagnetic resonance spectra for the series of the samples with different separations measured with the field applied transverse to the long axis and in the plane of the array is presented in Fig.3. The position of the main resonant peak shifts only slightly when the separation is reduced from 2000 to 1000 nm, however the shift becomes significant for separations smaller than 600 nm. Similar behavior was observed for the field applied along the stripes. Although the spectra in Fig. 4 are more complex, also in this case the resonant curves clearly shift towards higher fields for the arrays with smaller separations. This effect is especially pronounced when the separation decreases below 0.6 μm. It is important to note that in addition to the main resonance, due to coherent precession of the spins, additional resonances are visible in Figs.3 and 4. It is believed that they represent magnetostatic modes due to dimensional confinement of spinwaves in the stripes [8-10] FMR measurements for different directions of the field applied in the array plane are summarized in Fig. 5. The angular dependencies of the resonant field are

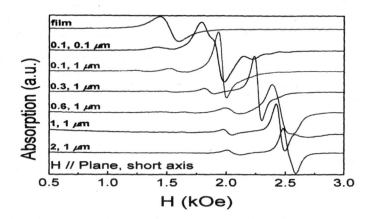

Fig. 3. Differential ferromagnetic resonance curves for 6 arrays with different separations (first number refers to the separation between long edges (in micrometers) and the second denotes the separation between rows). The top curve represents FMR signal of the reference film. The DC field was applied in the array plane and transverse to the stripes.

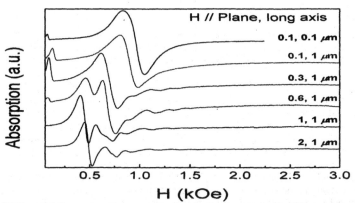

Fig. 4. Differential ferromagnetic resonance curves for 6 arrays with different separations (first number refers to the separation between long edges (in micrometers) and the second denotes the separation between rows). The DC field was applied in the array plane and along the stripes.

uniaxial anisotropy, the magnitude of which is determined by the extreme resonant fields: the lowest for the easy magnetization direction and the largest for the hard magnetization direction. The span of the resonant filed (vertical range of the curves) monotonically decreases as the stripes become closer to each other.

Reduced separations translate into increased strength of interactions. Thus, the effective uniaxial anisotropy of the arrays, resulting from the shape anisotropy of the individual stripes, decreases with the increasing strength of interactions. There was no angular variation of the in-plane resonant field observed for the reference sample of solid Py film (or no measurable in-plane anisotropy was present in the films)

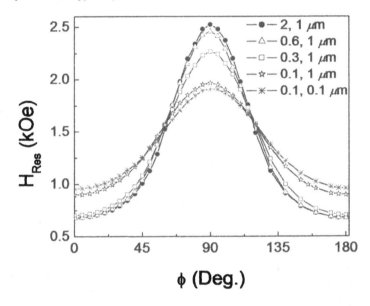

Fig. 5. Angular dependencies of the resonant fields for arrays with different separations between the stripes.

CONCLUSIONS

The results of measurements of magnetic hysteresis loops and the ferromagnetic resonance show that interactions among nanosized stripes in the patterned arrays have significant influence on their static and dynamic magnetic properties. The properties of the arrays with the largest separation of 2000 nm resemble those of individual stripes. However, as we reduce the separations between stripes the magnetostatic fields produced by individual stripes provide progressive coupling between them. Consequently, the properties of the arrays are no longer exclusively determined by isolated elements of the matrix but they involve collective effects of magnetostatic fields.

One of them is uniaxial magnetic anisotropy, which is markedly suppressed when the separation between particles decreases below 600 nm. Also, the coercive field is reduced when the strength of the coupling between neighboring stripes increases. The properties of the arrays with strong interactions evolve towards those of the continuous film and would eventually merge, however even for the separations as small as100 nm, distinct differences between the arrays and the reference film have been found. These separations seem to be large enough to prohibit propagation of domain walls in arrays, which can move freely in the solid film. Thus the magnetization mechanism in the arrays and the film are expected to be different for separations even smaller than 100 nm. Detailed understanding of these properties requires using micromagnetic modeling.

ACKNOWLEDGMENTS

This work was supported through the National Science Foundation under NIRT grant CCF-0403673 and through the DARPA grant HR0011-09-1-0047

REFERENCES

1. B. Kuanr, L. Malkinski, R.E. Camley, Z. Celinski, P. Kabos, *J. Appl. Phys* **93** 8591(2003)

2. M. Vroubel, Y. Zhuang, B. Rejael and J.N. Burghartz, *J. Appl. Phys.* **99** 08P506 (2006)

3. E. Salahun, P. Quéffélec, G. Tanné,A.-L. Adenot and O. Acher, *J. Appl. Phys.* **91** 5449(2002)

4. C. Han, B. Y. Zong, P. Luo, Y.H. Wu, *J. Appl. Phys.* **93** 9202 (2001)

5. A. Vovk, L. Malkinski, V. Golub, S. Whittenburg, C. O'Connor, J.-S. Jung, S-H. Min, *J. Appl. Phys* **97**,(10) 10J506 (p.1-3) (2005)

6. M. Yu, L. Malkinski, L. Spinu, W. Zhou, S. Whittenburg", *J. Appl. Phys.*101 09F501(2007)

7. L. Malkinski, M. Yu, A. Vovk, D.J. Scherer II, L. Spinu, W. Zhou, S. Whittenburg, Z. Davis, *J. Appl. Phys,* **101**, 09J110(2007)

8. S. O. Demokritov, B. Hillebrandts, A. N. Slavin. *IEEE Trans. Magn.* **38** 2502 (2002)

9. G. Gubbiotti, G. Carlotti, T. Okuno, T. Shinjo, F. Nizzoli and R. Zivieri, *Phys Rev B*, **69** 184409(2003)

10. R. Arias and D. L. Mills, Phys. Rev. B **70** 094414(2004)

Mater. Res. Soc. Symp. Proc. Vol. 1250 © 2010 Materials Research Society   1250-G08-04

Direct Synthesis of $L1_0$-Phase Nanostructured CoPt Thin Films Using Dense Plasma Focus Device Operating in Non-optimized Focus Mode

Zhenying Pan[1], Rajdeep S. Rawat[1], Jiaji Lin[2], Shumaila Karamat[1], Paul C.K. Lee[1], Stuart V. Springham[1], Augustine T.L. Tan[1],

[1]NSSE, National Institute of Education, Nanyang Technological University, 1 Nanyang Walk, Singapore 637616

[2]Solar Energy Research Institute of Singapore, National University of Singapore, 7 Engineering Drive 1, Singapore 117574

ABSTRACT

A direct synthesis of (001) oriented nanostructured CoPt thin films has been successfully achieved using a 880 J pulsed Dense Plasma Focus (DPF) device operating in a non-optimized focus mode with a low charging voltage of about 8 kV. The (001) oriented *fct* structured $L1_0$ phase nanostructured CoPt thin films have been synthesized directly in as-deposited sample, as verified by XRD results, without any post deposition annealing. The SEM imaging results show that nanostructured CoPt were achieved in non-optimized focus mode with agglomerate/particle size ranging from 10 to 55 nm. Furthermore, the VSM analysis shows that the as-deposited samples in non-optimized focus mode have higher coercivity (due to direct $L1_0$ phase) as compared the annealed sample and the as-deposited sample of optimized focus mode operation.

INTRODUCTION

The CoPt binary alloy is an excellent system because it has both chemical stability and high magnetocrystalline anisotropy which has received significant attention as possible high density magnetic recording media owing mainly to the existence of ordered intermetallic phases with exceptionally hard magnetic properties. The ordered *fct* structure $L1_0$ CoPt nanoparticles have a large magnetic anisotropy of about 4.9×10^7 erg/cc [1], hence the thermal agitation would be suppressed due to their large magnetic energy, which makes it a potentially interesting material from a commercial standpoint. From the practical application points of view, the magnetic data storage requires: (i) small particle size with tight size distribution; (ii) *fct* phase with high magnetocrystalline anisotropy and preferred (001) oriented crystal to overcome the superparamagnetism; and (iii) well-separated nanoparticles for reduced exchange coupling effects. Hence these are three key challenges for practical data storage applications of $L1_0$ phase CoPt nanoparticles nowadays.

CoPt nanoparticles have been produced using various techniques, such as chemical methods [2,3] and physical thin film deposition techniques including pulsed laser deposition (PLD) [4], the radio-frequency (RF) sputtering [5], molecular beam epitaxy (MBE) [6], etc. In our research work, we used a Dense Plasma Focus (DPF) device as a pulsed plasma deposition source for nanostructured $L1_0$-CoPt thin films synthesis. Being a source of a wide range of phenomena,

DPF has found applications in material science. Many different kinds of thin films have been deposited and processed using DPF [7-11]. It has been reported that the energetic ions of the DPF device can be used for nanostructuring of PLD synthesized FePt thin films and also for inducing a change of phase in magnetic thin films [12-14]. The DPF device has also been successfully employed for deposition of magnetic thin films and nanoparticles [9,10]. More recently, we have achieved nanostructured CoPt thin film synthesis using DPF as a pulsed plasma deposition source operating in an optimized focus mode, i.e. with best possible focusing efficiency. In this paper, we present the successful synthesis of the (001) oriented ordered Ll_0 phase magnetic CoPt thin films using DPF device, but operated in non-optimized mode.

EXPERIMENTAL SETUP

In present investigation, a high performance low inductance capacitor bank based repetitive DPF designated as NX2 (Nanyang X-ray source) was employed to synthesize the nanostructured CoPt thin films. The device was operated in non-optimized mode to avoid exposure of the deposited material to high energy ions which can affect their structure. The schematic of the deposition setup is shown in figure 1.

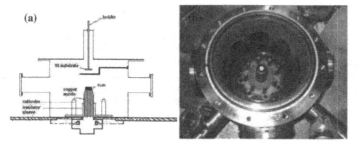

Figure 1 (a) The schematic of deposition setup and (b) top view of NX2 deposition chamber.

The conventional central hollow copper anode was replaced by a high purity (50:50 at%; Kurt J. Lesker, 99.99%) solid CoPt tip fitted copper anode. It may be interesting to mention that, in our case, for materials synthesis, the NX2 device, which has a 27.6 μF capacitor bank, was operated at much lower voltage of 8 kV with the stored energy of about 880 J.

To operate in a non-optimized mode, we adjusted the filling gas pressure just below the lowest possible focusing pressure, so that the efficient focusing/pinching of the plasma does not take place. The optimized pressure for hydrogen filled NX2 device is in the range from 1 to 30 mbar. Hence, for this experiment the filling gas pressure of hydrogen gas was kept at 0.50 mbar. The deposition distance was fixed at 10 cm which is much closer to the anode as compared with the distance used in the optimized focus mode, because the material ablation in non-optimized focus mode was expected to be less intense as compared to that of optimized focus mode. A shutter was used, between the anode and the sample holder, to avoid deposition during the initial three to five shots to verify that the device is working in desired mode. Once the system is working consistently in non-optimized mode, the shutter was removed for the designated number shots for thin film deposition.

The number of plasma focus deposition shots was fixed at 25 shots and the firing frequency was fixed at 1 Hz. Samples were consequently annealed at temperatures of 600 °C in vacuum for 1 hour duration to investigate the effects of annealing on the structural and magnetic properties of the deposited CoPt thin films.

The morphology was studied by JOEL JSM-6700F field emission scanning electron microscope (FESEM). X-ray diffraction (XRD) patterns, for structural properties analysis, were obtained by a Rigaku D/MAX-rA X-ray diffractometer with CuKα radiation. A Lakeshore 7400 vibrating sample magnetometer (VSM) was used to record hysteresis loops at room temperature to investigate the magnetic properties.

RESULTS AND DISCUSSIONS

Morphological features

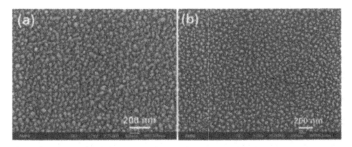

Figure 2 SEM images of CoPt samples synthesized in non-optimized focus mode at a filling gas pressure of 0.50 mbar: (a) the as-deposited sample, and (b) the sample after annealing at 600 °C for 1 hour in vacuum.

Figure 2 (a) & (b) show the morphological features of the as-deposited sample and corresponding annealed sample synthesized at a filling gas pressure of 0.50 mbar. It was observed that two types of agglomerates are formed on the sample, referring to figure 2(a), with a relatively broad size distribution. The bigger sized agglomerates are seen to grow from the smaller sized agglomerates by coalescing at the boundary of smaller particles/agglomerate or by coalescing with new incoming particles during deposition. The smaller sized agglomerates/particles can be noticed between big sized agglomerate and appear to form the background beneath the bigger sized agglomerates. The bigger agglomerates have an average size of about 51.9 ± 4.9 nm while the smaller agglomerates/particles have average size of 16.4 ± 4.0 nm. The morphological features of the sample after annealing, shown in figure (b), changed significantly. The agglomerates have changed their shape due to the sintering which causes atoms at surface or grain boundaries undergoing solid-state diffusion when it was annealed at elevated temperature of 600 °C. The small size agglomerates were found between the big agglomerates and no longer seems to form the background (as seen in the as-deposited sample before annealing). The average agglomerates size for both kinds of agglomerates after annealing changed slightly as compared with that of the sample before annealing, and are estimated to be

115

about 53.0 ± 4.6 nm and 20.2 ± 12.2 nm for the big and small agglomerates, respectively.

Structural properties

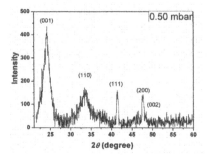

Figure 3 XRD patterns of the as-deposited CoPt samples synthesized at a filling gas pressure of 0.50 mbar in non-optimized focus mode.

Figure 3 shows the XRD patterns of the as-deposited CoPt samples synthesized in non-optimized focus mode. The most striking feature, in figure 3, is that (001)-oriented *fct* structured $L1_0$ phase CoPt nanostructures have directly been achieved on the as-deposited sample synthesized in non-optimized focus mode. It is well known that the distinct features for *fct* structured $L1_0$ phase are the appearance of the (001)/(110) superlattice peaks and the splitting of fundamental (200) peak into (200) and (002) superlattice peak due to the tetragonal lattice distortion [15]. In our case, (001) peaks is clearly observed at about 24.0° in the XRD patterns which indicates that a *fct* structured $L1_0$ phase CoPt nanostructures was directly achieved in non-optimized focus mode. This is interesting since the direct *fct* structured $L1_0$ phase has not been achieved up date on as-deposited samples using any physical deposition techniques without any substrate heating assistance. It has been reported that (001) texture were obtained by epitaxial growth on heated $MgO_{(001)}$ substrates [16]. Nearly perfect (001) textured CoPt and FePt films by directly sputtering films on thermally oxidized Si substrates and subsequent annealing [17]. Hence, it is worth to emphasize that the direct (001) oriented films achieved in non-optimized focus mode are technologically significant results since the c-axis is the easy axis of $L1_0$ phase CoPt and its orientation either in the plane or perpendicular to the plane is desirable for the application in data-storage.

The intensity ratio of the (001) superlattice peak to the (111) fundamental peak, i.e. $I_{(001)}/I_{(111)}$, reveals the (001)-orientation degree of the corresponding film [18]. Hence, orientation degree of as-deposited samples, in present case, could be estimated by the ratio of intensities of (001) and (111) peaks. The ratios of $I_{(001)}/I_{(111)}$ for the as-deposited samples synthesized in non-optimized focus mode are estimated to be about 8.94. The larger the ratio $I_{(001)}/I_{(111)}$, the better the (001) texture i.e. 'orientation degree'. Moreover, the orientation of (001) peak coincides with magnetic easy axis of the ordered $L1_0$ *fct* phase. Higher is the orientation degree of magnetic easy axis in the as-deposited *fct* phase CoPt thin films higher will be the magnetocrystalline anisotropy resulting in high coercivity.

The direct formation of (001) oriented *fct* structure $L1_0$ phase in as-deposited CoPt thin films in non-optimized focus mode can be attributed to the absence of ion-irradiation effect. In optimized focus mode, instability generated high energy ion beam (with ion energies ranging from few tens of keV to few MeV) processes the deposited material and these energetic ions may lead to destruction the crystal structure due to excessive energies. In non-optimized focus mode, the the pinch column and hence the plasma instabilities and thereafter the ion beam are not formed for over-processing of the deposited material and ablated anode material has just sufficient energy to go into the right phase, since the activation energy for ordering is only about few eV for CoPt alloy.

Magnetic properties

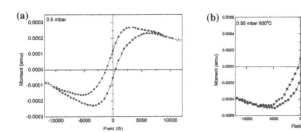

Figure 4 The hysteresis loops of the CoPt samples synthesized in non-optimized focus mode at a filling gas pressure of 0.50 mbar (a) the as-deposited sample and (b) the sample after annealing at 600 °C for 1 hour.

The hysteresis loop of as-deposited samples synthesized at a filling gas pressure of 0.50 mbar in non-optimized focus mode is shown in the figure 4(a) and the corresponding sample after annealing at 600 °C in vacuum for 1 hour is shown in the figure 4(b). It was observed the as-deposited samples show harder magnetic property with a coercivity close to 1000 Oe. This indicates that a self ordering process has taken place in the as-deposited sample when the filling gas pressure was proper for the deposition. This is consistent with the XRD results as the intensity of the (001) superlattice peak is stronger compared to that of fundamental peak (111). The coercivity value is 838 Oe for as-deposited sample. The coercivity drops down to 566 Oe for the sample annealed for 1 hour in vacuum at 600 °C. This deterioration of magnetic properties may be caused by the increase of exchanged coupling of the CoPt nanostructures as the agglomerates on the annealed samples are closely packed, as seen in figure 2. The as-deposited samples of non-optimized focus mode are magnetically harder than the as-deposited samples of the optimized focus mode depositions, which were usually much less than 100 Oe as reported in our earlier research work [19]. This therefore confirms the direct formation of the magnetically hard *fct* structured $L1_0$ phase on the as-deposited sample synthesized in non-optimized focus mode.

CONCLUSIONS

The nanostructured CoPt thin films are successfully synthesized by NX2 DPF device operating in a non-optimized focus mode with hydrogen filling gas pressure of 0.50 mbar at a relatively close distance of 10 cm from the anode top. The XRD results confirmed the direct synthesis of (001) oriented fct structured $L1_0$ phase nanostructured CoPt thin films in as-deposited sample without any post annealing. The morphological features observed in SEM images show that nanostructured CoPt thin films in the form of nanoparticles or agglomerates were achieved with the feature size in the range of 10 to 55 nm. Furthermore, the VSM analysis shows that the as-deposited samples have higher coercivity as compared with the sample after annealing and the as-deposited sample synthesized in optimized focus mode operation.

REFERENCES

1   D. Weller, A. Moser, L. Folks, M. E. Best, W. Lee, M. F. Toney, M. Schwickert, J. U. Thiele, and M. F. Doerner, IEEE Trans. Magn. **36**, 10-15 (2000).
2   S. H. Sun, C. B. Murray, D. Weller, L. Folks, and A. Moser, Sci **287**, 1989-1992 (2000).
3   X. C. Sun, Z. Y. Jia, Y. H. Huang, J. W. Harrell, D. E. Nikles, K. Sun, and L. M. Wang, J. Appl. Phys. **95**, 6747-6749 (2004).
4   E. Agostinellia, S. Lauretia, G. Varvaroa, A. Generosib, B. Pacib, V. Rossi-Albertinib, G. Scaviaa, and A. M. Testaa, Mater. Sci. Eng., C **27**, 1466-1469 (2006).
5   Hu, J.-P., and P. Lin, IEEE Trans. Magn. **32**, 4096 - 4098 (1996).
6   W. Grance, C. Ulhaq-Bouillet, M. Maret, and J. Thibault, Acta Mater. **49**, 1439–1444 (2001).
7   R. S. Rawat, W. M. Chew, P. Lee, T. White, and S. Lee, Surf. Coat. Tech. **173**, 276 (2003).
8   L. Y. Soh, P. Lee, X. Shuyan, S. Lee, and R. S. Rawat, IEEE Trans Plasma Sci **32**, 448-455 (2004).
9   R. S. Rawat, T. Zhang, K. S. T. Gan, P. Lee, and R. V. Ramanujan, Appl. Surf. Sci. **253**, 1611-1615 (2006).
10   T. Zhang, K. S. Thomas. Gan, P. Lee, R. V. Ramanujan, and R. S. Rawat, J. Phys. D: Appl. Phys. **39**, 2212-2219 (2006).
11   R. S. Rawat, P. Arun, A. G. Vedeshwar, P. Lee, and S. Lee, J. Appl. Phys. **95**, 7725-7730 (2004).
12   J. J. Lin, M. V. Roshan, Z. Y. Pan, R. Verma, P. Lee, S. V. Springham, T. L. Tan, and R. S. Rawat, J. Phys. D: Appl. Phys. **41**, 135213 (2008).
13   Z. Y. Pan, J. J. Lin, T. Zhang, S. Karamat , T. L. Tan, P. Lee, S. V. Springham, R. V. Ramanujan, and R. S. Rawat, Thin Solid Films **517**, 2753–2757 (2009).
14   Z. Y. Pan, R. S. Rawat, J. J. Lin, T. Zhang, P. Lee, T. L. Tan, and S. V. Springham, Appl Phys a-Mater **96**, 1027-1033 (2009).
15   C. L.Platt, K. W.Wierman, E. B.Svedberg, R. v. d. Veerdonk, J. K.Howard, A. G.Roy, and D. E.Laughlin, J. Appl. Phys. **92**, 6104-6109 (2002).
16   B. M. Lairson, M. R. Visokay, R. Sinclair, and B. M. Clemens, Appl. Phys. Lett. **62**, 639 (1993).
17   H. Zeng, M. L. Yan, N. Powers, and D. J. Sellmyer, Appl. Phys. Lett. **80**, 2350-2352 (2002).
18   C. Chen, O. Kitakami, S. Okamoto, and Y. Shimada, Appl. Phys. Lett. **76**, 3218-3220 (2000).
19   Z. Y. Pan, R. S. Rawat, M. V. Roshan, J. J. Lin, R. Verma, P. Lee, T. L. Tan, and S. V. Springham, J. Phys. D: Appl. Phys. **42**, 175001 (2009).

Ferroelectric RAM (FeRAM)

Mater. Res. Soc. Symp. Proc. Vol. 1250 © 2010 Materials Research Society   1250-G11-01

Overview and Technical Trend of Chain FeRAM

Daisaburo Takashima
Center for Semiconductor Research & Development, Semiconductor Company, Toshiba Corp.,
2-5-1, Kasama, Sakae-ku, Yokohama 247-8585, Japan

ABSTRACT

A chain FeRAM^{TM} is the best solution to realize high-speed and high-bandwidth nonvolatile RAM with low power dissipation. In this paper, the overview of chain FeRAM, the technical trend for FeRAM scaling and the marketing strategy are presented. First of all, the concept and performance of chain FeRAM are described. Secondly, the status and history of chain FeRAM development are presented. Thirdly, four kinds of scaling strategies for chain FeRAM are presented; (1) A shrink trend of chain cell including a capacitor plug shared with twin cells, and process techniques including Ir/TiAlN-barrier metal and MOCVD-PZT with $SrRuO_3$ electrode, which are installed in 16Kb, 8Mb, 32Mb, 64Mb and 128Mb chain FeRAMs, (2) Capacitor damage suppression processes to reinforce step coverage and protect H_2 damage even in $0.1\mu m^2$ capacitor of 128Mb, (3) A scalable array architecture such as an octal / quad bitline architecture to reduce bitline capacitance and ensure enough cell signal in scaled ferroelectric capacitor, and (4) A ferroelectric capacitor overdrive technique by driving shield-bitlines to enlarge tail-to-tail cell signal in low voltage operation of 1.3V. Fourthly, future direction of chain FeRAM is discussed. The vertical capacitor is one of candidates for gigabit-scale chain FeRAMs, and solves signal problem and achieves small $4F^2$ cell without contact formation. Finally, the marketing strategy to take full advantage of chain FeRAM is presented. A nonvolatile FeRAM cache is the promising candidate to achieve high bandwidth memory systems. Applications of chain FeRAM to solid-state drive (SSD) and hard-disk drive (HDD) and their system performance improvements are demonstrated.

INTRODUCTION

A ferroelectric random access memory (FeRAM) using polarization switching of ferroelectric film is applicable to many kinds of portable electronic equipments such as smart card, cellar phone and nonvolatile cache in memory systems, because the FeRAM provides excellent characteristics such as non-volatility, low power consumption, fast read/write operation and high endurance of over 10^{13} read/write cycles. In the late 1980s, early papers on 512b and 16-kb FeRAMs [1-2] with 2 transistors and 2 capacitors (2T/2C) cell were reported. The first 256-kb FeRAM using 1T/1C cell was demonstrated in 1994 [3]. Since then, many efforts to shrink memory cell size and die size have been made. These efforts have realized 64Mb FeRAMs [4-5] as shown in Fig. 1. These memory cells consist of 1-transistor and 1-capacitor in series, and adopt the cell structure with capacitor-under-bitline [4] or capacitor-over-bitline [5] like DRAM. However, these conventional FeRAMs have some drawbacks such as (1) the cell size limitation to $8F^2$, (2) slow plateline drive and large plateline driver size due to large load of ferroelectric capacitors, and (3) small cell signal due to large parasitic bitline capacitance. On the other hand, a chain FeRAM architecture has been proposed in 1997 [6]. The chain FeRAM overcomes these problems and realizes small cell / die size, fast plateline drive with small plateline driver and large cell signal even in scaled 128Mb.

In this paper, the overview of chain FeRAM and technical trend are presented. First, the concept of chain FeRAM is presented. Second, the development history of chain FeRAM is presented. Third, the scaling techniques in many aspects and the future direction for chain FeRAM are discussed. Finally, the market strategy to take full advantage of chain FeRAM is presented.

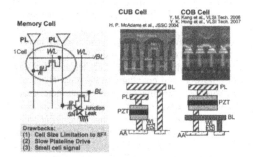

Figure 1. Configuration and problem of the conventional FeRAM.

CONCEPT OF CHAIN FERAM

The concept of chain FeRAM architecture [6] is shown in Fig. 2. The chain FeRAM realizes smallest $4F^2$ size memory cell in ideal case using a planar cell transistor due to cross-point-type cell, and achieves fast random access. A memory cell consists of 1-transistor and 1-ferroelectric capacitor connected in parallel. A chain cell string consists of plural memory cells connected in series. One terminal of the chain cell string is connected to bitline BL via a block selector, and the other terminal is connected to a plateline PL. One PL is shared with plural memory cells. This enables large PL driver with small total PL driver area. A BL-contact to memory cells is also shared with plural memory cells. This reduces BL capacitance per cell to 1/2 to 1/3 of that of the conventional FeRAM, and achieves large readout cell signal due to small total BL capacitance.

Fig 3 shows basic operation of chain FeRAM. During standby cycle of Fig. 3 (a), all wordlines WL_0-WL_3 are boosted to high voltage, and the block selecting signal BS and the PL are grounded. Therefore all ferroelectric capacitors are short-circuited by cell transistors turned-on as shown in the hysteresis loop, and both "1" and "0" data are held steadily at zero bias points without noise from capacitor electrodes. During active cycle of Fig. 3 (b), when accessing data of the cell in a chain cell string, the selected wordline WL_2 is pulled down and the BS is pulled up. After that, the PL is pulled up to array operating voltage Vint of 1.8V-1.5V. Therefore, the PL

122

bias is applied only to the selected ferroelectric capacitor, and the cell data is read out to BL, as shown in the loci of the hysteresis loop. When "1" data is read out, a large signal charge is read out to BL by polarization switching. When "0" data is read out, a small signal charge is read out to BL by non-polarization switching. After that, a sense amplifier amplifies the readout cell signal. On the other hand, cell data of the unselected cells are protected by short-circuiting capacitors. Thus, a complete random access is realized.

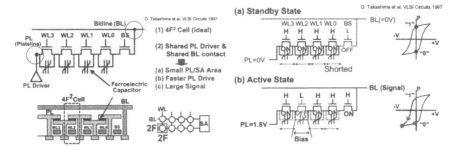

Figure 2. Concept of chain FeRAM.   **Figure 3.** Basic operation of chain FeRAM.

Figure 4 shows the chip size and readout cell signal comparisons between the conventional FeRAM and chain FeRAM. As shown in the 128Mb chip breakdown of Fig. 4 (a), the conventional FeRAM needs large PL driver area for driving many ferroelectric capacitors. On the other hand, PL driver area in the chain FeRAM can be reduced to 1/3 of that of the conventional FeRAM by sharing PL with plural chain cells. Moreover, in 128Mb-generation, a memory cell size is limited by front-end-of-line (FEOL) rule. This realizes small chain cell size, compared with that of the conventional cell. As a result, the chip reduction by 30% is achieved by introducing the chain FeRAM architecture. As shown in Fig. 4 (b), in 64Mb-generations, the chain FeRAM realizes twice cell signal than that of the conventional FeRAM, because BL capacitance per cell in chain FeRAM is small by sharing BL-contact with plural chain cells. This enables enough cell signal of ±200mV even in scaled 128Mb chain FeRAM by introducing scalable array architecture discussed later.

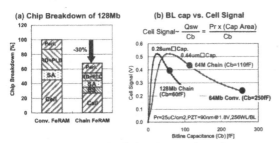

Figure 4. Chip size and cell signal comparisons between the conventional FeRAM and chain FeRAM.

DEVELOPMENT STATUS OF CHAIN FERAM

The status and history of chain FeRAM development are summarized in Fig. 5. The first prototype of 16Kb chain FeRAM [7] has been developed in 1999. Since then, 8Mb [8] in 2001, 32Mb [9] in 2003, 64Mb [10] in 2006 and 128Mb [11] in 2009 have been developed so far. The average cell size including a block selector region has been shrunk from $13\mu m^2$ to $0.32\mu m^2$. A random cycle time for each chip is around 70ns, whereas the page/burst cycle time has been drastically reduced from 80ns to 1.25ns. The latest 128Mb chain FeRAM has the highest memory capacity in all nonvolatile RAMs. The die size is 87.7 mm^2. This chip adopts DDR2 SDRAM interface of highest read/write bandwidth of 1.6GB/s (=16IO / 1.25ns) in all nonvolatile memories at 400MHz clock and 1.8V Vdd.

Design Rule	0.5µm	0.25µm	0.2µm	130nm	130nm +DualOx
Capacity	16Kb	8Mb	32Mb	64Mb	128Mb
Chip	 ISSCC1999	 ISSCC2001	 ISSCC2003	ISSCC2006	 ISSCC2009
Cell Structure	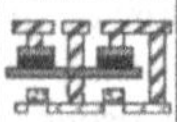 Offset Cell	 Offset Cell	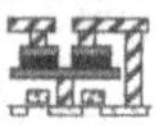 Stacked Cell	 1-Mask Cell	 Non-separated PZT film
Cell Size(Ave.)	13.26µm²	5.2µm²	1.875µm²	0.7191µm²	0.32µm²
Chip Size	1.9mm²	76mm²	96mm²	87.5mm²	87.7mm²
R/W Cycle	80ns	70ns	75ns	60ns	75ns
Page/Burst	80ns	40ns	25ns	10ns	1.25ns

Figure 5. Development history of chain FeRAMs.

SCALING STRATEGY

In this session, 4 kinds of scaling techniques; (1) memory cell designs and materials to minimize cell size, (2) capacitor damage suppression schemes to obtain sufficient cell signal, (3) an scalable array architecture to reduce BL capacitance, and (4) a ferroelectric capacitor overdrive technique to achieve low-voltage operation are discussed.

Memory Cell Design and Technology

Figure 6 shows trend of chain memory cell. In the offset cell of 8Mb, a bottom electrode was connected to an active area via M1 metal. This caused large memory cell size. Since 32Mb, the stacked cell technology has been introduced to connect the bottom electrode directly to the active area by a capacitor plug. An Ir/TiAlN/W-plug structure [12-15] prevents plug-oxidation even at 650C 1-hour. This is enough for PZT crystallization annealing. In 32Mb, the bottom electrode is shared with neighboring twin cells, utilizing a merit of chain cell in order to reduce cell size [12]. In 64 Mb, one mask cell was introduced to reduce process cost and minimize taper of capacitor [13-14]. In 128Mb, A design rule of front-end-of-line (FEOL) process limits memory cell size. Therefore, the bottom electrode and the capacitor plug are shared with twin cells again [15]. A MOCVD-PZT adopted since 64Mb [14-15] has enabled high-density PZT film and guarantees 10 year data retention at 85C. A $SrRuO_3$ top electrode adopted since 8Mb have given excellent fatigue property. Furthermore, in order to maximize the effective ferroelectric capacitor area, a triangular capacitor cell in addition to ArF lithography has been installed in 128Mb as shown in Fig.7. This enlarges capacitor area by 17%.

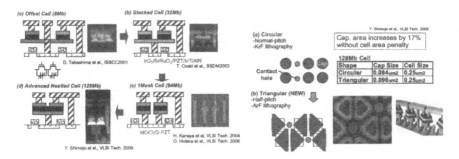

Figure 6. Trend of chain cell development.   **Figure 7.** Capacitor design in 128Mb.

Capacitor Damage Suppression

A key process / integration issue for FeRAM is how to prevent hydrogen damage from back-end-of-line (BEOL) process. One solution is the improvement of step coverage of film protecting PZT from H_2 damage. Figure 8 shows an advanced chain cell structure for 128Mb. The ferroelectric film as well as bottom electrode is shared with twin cells [15]. This makes it easy to fill the protection film in a gap between capacitors and improves step coverage in addition to less PZT RIE damage. The cell signal improvement by 18% has been obtained. The other is the improvement of quality of protection film. Figure 9 shows the comparison between high-density and low-density cover films. High-density cover film protects H_2 damage from BEOL H_2 process [15]. No signal degradation has been observed even in $0.1\mu m^2$ capacitor.

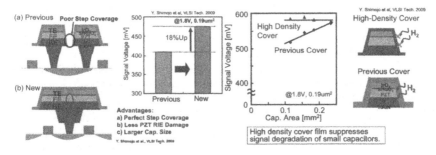

Figure 8. Advanced chain cell structure.   **Figure 9.** High-density H_2-block cover film.

Scalable Array Architecture

A serious problem of FeRAM scaling is the cell signal degradation due to ferroelectric capacitor size shrink using a planer capacitor. This is because a cell signal is proportional to ratio of cell polarization P_{cell} to BL capacitance Cb; (P_{cell}/Cb), and the Cb can not be scaled sufficiently in proportion to ferroelectric capacitor area shrink. Although chain FeRAM can reduce total BL capacitance, this problem is severe in 64Mb and beyond. Therefore, the scalable array architectures such as a quad BL architecture [10] and an octal BL architecture [11] have been introduced in 64Mb and 128Mb chain FeRAMs as shown in Fig. 10. These architectures reduce total BL capacitance to about 1/2, 1/4 by reducing total BL length, without increasing a sense amplifier area overhead. This is because the quad/octal BL architecture enables cell data accessing to one of four / eight BLs, and a sense amplifier is shared with four /eight bitlines. The octal BL architecture reduces BL capacitance to 60 fF and realizes ±200mV cell signal. A concern of accessing cell data to one of 8 BLs is area overhead to arrange 8 kinds of platelines PL_{0-7} to select one of 8 chain cell strings. This concern is solved by introducing 4 kinds of cell string lengths of 6, 7, 9 and 10 as shown in Fig. 11. Eight platelines PL_{0-7} are arranged over chain cell strings. The different contact positions, which are obtained by different chain string lengths, enable easy contacts between PLs and ends of chain cell strings.

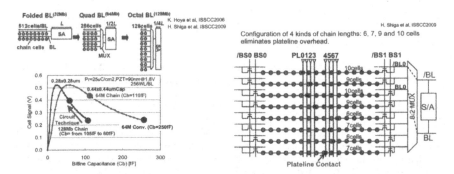

Figure 10. Scalable array architecture in 64Mb and 128Mb.

Figure 11. Octal-bitline architecture to select one of 8 chain cell strings.

Overdrive Technique

Another concern of FeRAM is the cell signal degradation due to low-voltage operation. In FeRAM operation, the polarization is switched when applying bias larger than coercive voltage Vc. Therefore, the maximum cell signal of FeRAM is approximately expressed as Vint-Vc, where the Vint is array operating voltage. The difficulty of Vc scaling due to ferroelectric capacitor leak problem causes signal degradation at low voltage operation. Figure 12 shows the measured cell signal of 576Kb Array TEG fabricated using 64Mb chain FeRAM process [16]. The capacitor size is 0.44μm x 0.44μm. The decrease of array voltage Vint from 1.7V to 1.3V causes the peak-to-peak cell signal degradation from 520mV to 350mV. Furthermore, the cell signal distribution worsens tail-to-tail cell signal from 260mV to 90mV. In order to avoid cell signal degradation due to low-voltage operation, a ferroelectric capacitor overdrive with shield-BL drive, called the shield-BL-overdrive technique [16], has been proposed in 2010 as shown in Fig. 13. The shield-BL-overdrive of Fig. 13 applies larger bias to ferroelectric capacitor without the increase of BL/PL swing range and total BL capacitance, and obtains large cell signal. In this technique, the shield-BLs are pulled down just when BL rises from ground level after the initial bias is applied to ferroelectric capacitor. Therefore, by the parasitic BL-to-BL coupling, the further bias of 0.24V is applied to the ferroelectric capacitor.

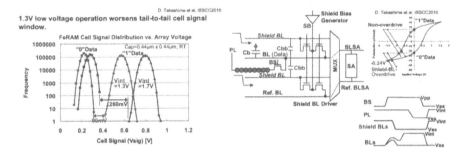

Figure 12. Problem of low-voltage operation in scaled chain FeRAM.

Figure 13. Shield-bitline-overdrive technique.

Fig. 14 shows the measured cell signal distribution of the 576Kb test chip at array voltage Vint of 1.5V. The peak-to-peak cell signal of 430mV in the non-overdrive has been improved to 530mV in the proposed technique, and the tail-to-tail signal of 190mV has also been improved to 300mV by 0.24V overdrive. Fig. 15 shows the measured dependency of the peak-to-peak and tail-to-tail cell signals on the array voltage Vint. The proposed technique has enhanced signal by 100mV at any Vint conditions. The tail-to-tail cell signal window (= 2 x 1T1C signal) has doubled and increased from 90mV to 180mV at low voltage operation of 1.3V. The signal equivalent to 1.5V operation has been obtained at 1.3V operation. The advantage of this technique is scalability. The overdrive voltage will increases in device scaling thanks to BL-pitch scaling. The overdrive voltage of 0.36V is expected in 512Mb-generation. It is expected that the signal of 256Mb / 512Mb will improve from 95mV / 60mV to 230mV / 200mV even in lowered supply voltage of 1.3V / 1.2V by introducing the proposed shield-BL-overdrive technique [16].

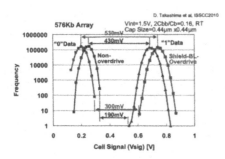

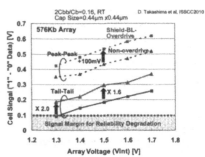

Figure 14. Measured cell signal improvement by shield-BL-overdrive.

Figure 15. Measured dependency of cell signal on array voltage.

FUTURE DIRECTION

In this session, the future direction to realize gigabit-scale FeRAMs is addressed. One approach to increase an amount of memory cell polarization and realize enough cell signals even in small cell is to introduce the cell with three-dimensional (3D) ferroelectric capacitor. Fig. 16 (a) shows 3D-capacitor applicable for the conventional COB-type cell [17]. A challenging issue to make FeRAM with 3D-capacitor is how to control the orientation of polarization over non-flat bottom electrode, because ferroelectric film has highly orientated polycrystalline structure. Another problem of the 3D ferroelectric capacitor in the conventional FeRAM is that the memory capacitor size is limited to 6F×6F, as shown in Fig. 16 (a), where F means each material thickness. The vertical capacitor [18] applicable to chain FeRAM of Fig. 16 (b) overcomes these problems and achieves small capacitor of 2F×2F size. This vertical capacitor utilizes the polarization orientation only in horizontal direction thanks to simple chain FeRAM structure. Moreover, the fabricated ferroelectric capacitor has good symmetrical property as shown in the hysteresis curve, because both electrodes are symmetrical and fabricated at a time.

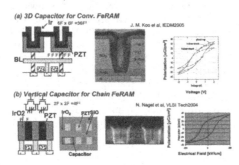

Figure 16. Future cell prospect of FeRAM.

MARKETING STRATEGY

In this section, a new application suitable for FeRAM is demonstrated. The most advantage of FeRAM is low write-energy per bit. Figure 17 shows trend of write-energy per bit in all emerging nonvolatile RAMs. In nonvolatile RAMs such as PRAM and MRAM, the cell data is written by flowing the current into the cell. This needs relatively large write-current per cell of 50µA to 2mA. On the other hand, in the chain FeRAM, the current to switch polarization of only 4µA to 13µA is required. This needs one less digit of write-energy per bit than others. This excellent property enables mass data read/write and achieves high read/write bandwidth with long page length. This realizes DRAM replacement with nonvolatile chain FeRAM. In order to take full advantage of chain FeRAM, a 128Mb chain FeRAM with 1.6GB/s DDR2 SDRAM interface has been developed [11]. The read/write bandwidth of 1.6GB/s is the highest in all nonvolatile memories reported so far as shown in Fig. 18.

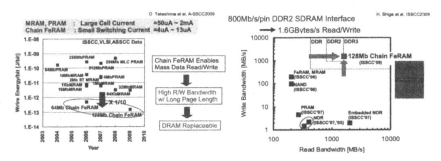

Figure 17. Trend of write-energy per bit in all emerging memories.

Figure 18. Read/write bandwidth comparison among all nonvolatile Memories.

One good application of FeRAM utilizing the merit of high-bandwidth is to use it as a nonvolatile FeRAM cache for solid-state drive (SSD) and hard-disk drive (HDD). Fig. 19 shows the concept of HDD/SSD with a nonvolatile FeRAM cache [19-20]. The windows OS such as Windows XP, Vista and 7 installed in a personal computer (PC) issues the flush cache commands at short intervals. This flush cache command requests HDD/SSD to move DRAM cache data into a track on disk / NAND flash memory to avoid data destruction at sudden power failure, because DRAM data is destroyed at power-off. Due to this command issue, as shown in Fig. 20, an average of valid write-data in DRAM cache is only 180KB when a business user uses PC for 22 days [19-20]. Therefore, a few valid data in DRAM cache not only degrades cache hit rate, but also causes a great deal of program access with fragmented data to the disk / NAND flash memory. On the other hand, in the nonvolatile FeRAM cache, the flush cache command can be ignored. Therefore, the 16MB FeRAM cache can be filled with valid write-data. This not only improves cache hit rate, but also achieves efficient data write-back. A lot of data can be written back into the disk / NAND flash memory at a time.

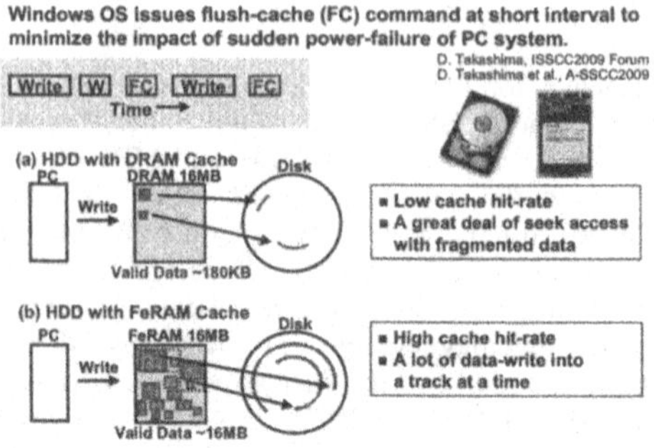

D. Takashima, ISSCC2009 Forum
D. Takashima et al., A-SSCC2009

Case: Business Use (User Data for 22 Days)

Valid write-data size in DRAM is only 180KB.

Write | W | FC | Write | FC
Time →

(a) HDD with DRAM Cache
PC   DRAM 16MB   Disk
Write
Valid Data ~180KB

- Low cache hit-rate
- A great deal of seek access with fragmented data

(b) HDD with FeRAM Cache
PC   FeRAM 16MB   Disk
Write
Valid Data ~16MB

- High cache hit-rate
- A lot of data-write into a track at a time

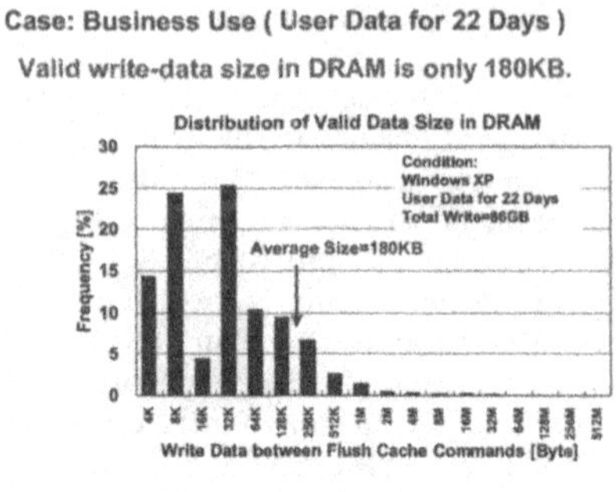

Distribution of Valid Data Size in DRAM

Condition:
Windows XP
User Data for 22 Days
Total Write=86GB

Average Size=180KB

Frequency [%]

Write Data between Flush Cache Commands [Byte]

Figure 19. Concept of FeRAM cache for HDD/SSD. **Figure 20.** Measured valid write-data size in DRAM cache of HDD.

Figure 21 shows the simulated effective write bandwidth of 2.5-inch HDD, where the measured PC user data for 5 days is used for simulation. In DRAM cache, valid write-data in cache is limited to 180KB. Therefore, the effective write bandwidth is low even in large cache memory capacity. On the other hand, in FeRAM cache, the effective write bandwidth increases with the increase of cache memory capacity. The write bandwidth when using a 128Mb chain FeRAM (16MB) is improved to 3.5 times of than that of DRAM [20]. Figure 22 shows a photograph of the prototype SSD with 16MB chain FeRAM cache and the measured score of PC Mark05, which is popular PC bench mark test. The X-axis shows the test items. The Y-axis shows the effective read/write bandwidth improvement ratio when using FeRAM cache. The FeRAM cache doubles read/write bandwidth in the XP startup, application loading, and general usage treating relatively small files. The total score has been improved to 1.5 times of that of DRAM cache [19].

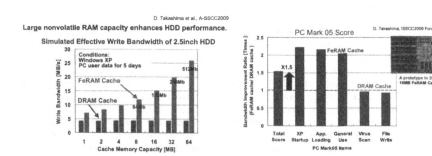

D. Takashima et al., A-SSCC2009

Large nonvolatile RAM capacity enhances HDD performance.

Simulated Effective Write Bandwidth of 2.5inch HDD

Conditions:
Windows XP
PC user data for 5 days

FeRAM Cache
DRAM Cache

Write Bandwidth [MB/s]

Cache Memory Capacity [MB]

PC Mark 05 Score

D. Takashima, ISSCC2009 Forum

X1.5

FeRAM Cache
DRAM Cache

A prototype in 2006:
16MB FeRAM Cache

Bandwidth Improvement Ratio [Times]
(FeRAM cache/ DRAM cache)

Total   XP   App.   General Virus File
Score Startup Loading Use   Scan   Write
PC Mark05 items

Figure 21. Simulated performance of HDD with FeRAM cache. **Figure 22.** Measured performance of SSD with FeRAM cache.

CONCLUSIONS

The overview of chain FeRAM, the technical trend for chain FeRAM scaling and the marketing strategy are presented. The chain FeRAM achieves small die by sharing PL-driver with plural cells and by small cross-point-type cell, and realizes large cell signal by small bitline capacitance due to sharing BL-contact with plural cells. The several key techniques used for 5-generations of chain FeRAMs; 16Kb, 8Mb, 32Mb, 64Mb and 128Mb are introduced. The advanced cell structure with a capacitor plug shared with twin cells and with the high-density cover film shrinks cell size to $0.25\mu m^2$, and enhances step coverage and protects H_2 damage. The scalable array architecture obtains enough cell signal even in small ferroelectric capacitor by minimizing total BL capacitance. The shield-BL-overdrive technique improves tail-to-tail cell signal at low-voltage operation of 1.3V. Furthermore, FeRAM scaling using a vertical capacitor to realize $4F^2$ size is discussed. Finally, the FeRAM cache application for HDD/SSD to take full advantage of chain FeRAM is demonstrated. The effective write bandwidth improvements have been verified by simulation and measurement.

REFERENCES

[1] J. T. Evans and R. Womack, "An experimental 512-bit nonvolatile memory with ferroelectric storage cell," IEEE J. Solid-State Circuits, vol. 23, no.5, pp. 1171-1175, Oct. 1989.

[2] R. Womack et al., "A 16kb ferroelectric nonvolatile memory with a bit parallel architecture," in IEEE ISSCC Dig. Tech. Papers, Feb. 1989, pp. 242-243.

[3] T. Sumi et al., "A 256kb nonvolatile ferroelectric memory at 3V and 100ns," in IEEE ISSCC Dig. Tech. Papers, Feb. 1994, pp. 268-269.

[4] H. P. McAdams et al., "A 64-Mb embedded FRAM utilizing a 130-nm 5LM Cu/FSG logic process," IEEE J. Solid-Sate Circuits, vol. 39, no. 4, pp. 1625-1634, pp. 667-677, Apr. 2004.

[5] Y. K. Hong et al., "130 nm-technology, 0.25 μm^2, 1T1C FRAM cell for SoC (system-on-a-chip)-friendly applications," in Symposium on VLSI Technology Dig. Tech. Papers, June 2007, pp. 230-231.

[6] D. Takashima et al., "High-density chain ferroelectric random access memory (chain FRAM)," IEEE J. Solid-State Circuits, vol. 33, no. 5, pp. 787-792, May 1998.

[7] D. Takashima et al., "A sub-40ns chain FRAM architecture with 7ns cell-plateline drive," IEEE J. Solid-State Circuits, vol. 34, no. 11, pp. 1557-1563, Nov. 1999.

[8] D. Takashima et al., "A 76-mm^2 8-Mb chain ferroelectric memory," IEEE J. Solid-State Circuits, vol. 36, no. 11, pp. 1713-1720, Nov. 2001.

[9] S. Shiratake et al., "A 32-Mb chain FeRAM with segment/stitch array architecture," IEEE J. Solid-State Circuits, vol. 38, no. 11, pp. 1911-1919, Nov. 2003.

[10] K. Hoya et al., "A 64Mb chain FeRAM with quad BL architecture and 200MB/s burst mode," in IEEE ISSCC Dig. Tech. Papers, Feb. 2006, pp. 134-135.

[11] H. Shiga et al., "A 1.6GB/s DDR2 128Mb chain FeRAM with scalable octal bitline and sensing schemes," IEEE J. Solid-State Circuits, vol. 45, no.1, pp. 142-152, Jan. 2010.

[12] T. Ozaki et al., "Key technologies of first chain – 32Mbit ferroelectric RAM," in Extended Abstracts of SSDM, Sep. 2003, pp. 646-647.

[13] H. Kanaya et al., "A 0.602μm^2 nestled chain cell structure formed by one mask etching process for 64Mbit FeRAM," in Symposium on VLSI Tech. Dig. Tech. Papers, June 2004, pp. 150-151.

[14] O. Hidaka et al., "High density high reliability chain FeRAM with damage-robust MOCVD-PZT capacitor with SrRuO$_3$/IrO$_2$ top electrode for 64Mb and beyond," in Symposium on VLSI Technology Dig. Tech. Papers, June 2006, pp.126-127.

[15] Y. Shimojo et al., "High-density and high-speed 128Mb chain FeRAM with SDRAM-Compatible DDR2 Interface," in Symposium on VLSI Tech. Dig. Tech. Papers, June 2009, pp.218-219.

[16] D. Takashima et al., "A Scalable shield-bitline-overdrive technique for 1.3V chain FeRAM, " in IEEE ISSCC Dig. Tech. Papers, Feb. 2010, pp. 262-263.

[17] J. M. Koo et al., "Fabrication of 3D trench PZT capacitors for 256Mbit FRAM device application," in IEEE IEDM Dig. Tech. Papers, Dec. 2005, pp. 340-343.

[18] N. Nagel et al., "New highly scalable 3-demential chain FeRAM cell with vertical capacitor," in Symposium on VLSI Technology Dig. Tech. Papers, June 2004, pp. 146-147.

[19] D. Takashima, "NAND flash memory and system designs for SSD application," in IEEE ISSCC Forum of "SSD Memory Subsystem Innovation", Feb. 2009.

[20] D. Takashima et al., "A 128Mb chain FeRAM and system designs for HDD application and Enhanced HDD performance", in IEEE A-SSCC Dig. Tech. Papers, Nov. 2009, pp. 13-16.

Mater. Res. Soc. Symp. Proc. Vol. 1250 © 2010 Materials Research Society   1250-G13-06

A Ferroelectric NAND Flash Memory for Low-Power and Highly Reliable Enterprise SSDs and a Ferroelectric 6T-SRAM for 0.5V Low-Power CPU and SoC

Kousuke Miyaji, Teruyoshi Hatanaka, Shuhei Tanakamaru, Ryoji Yajima, Shinji Noda, Mitsue Takahashi*, Shigeki Sakai* and Ken Takeuchi
Dept. of Electrical Engineering and Information Systems, University of Tokyo, Tokyo, Japan
*National Institute of Advanced Industrial Science and Technology, Tsukuba, Japan

ABSTRACT

This paper overviews recent research results about ferroelectric FETs (Fe-FETs) such as a ferroelectric (Fe-) NAND flash memory for enterprise solid-state drives (SSDs) and a ferroelectric 6T-SRAM for 0.5V operation low-power CPU and system-on-a-chip (SoC).

The Fe-NAND flash memory with a non-volatile (NV) page buffer is proposed. The data fragmentation of SSD in a random write is removed by introducing a batch write algorithm. As a result, the SSD performance can double. The NV-page buffer realizes a power outage immune highly reliable operation. In addition, a zero V_{TH} memory cell scheme is proposed to best optimize the reliability of the Fe-NAND. The V_{TH} shift caused by the read disturb, program disturb and data retention decreases by 32%, 24% and 10%, respectively. A 1.2V operation adaptive charge pump circuit for the low voltage and low power Fe-NAND is introduced. By using Fe-FETs as diodes in the charge pump and optimizing the V_{TH} of Fe-FETs at each pump stage, the power efficiency and the output voltage increase by 143% and 25% without the circuit area and process step penalty. Finally, a ferroelectric 6T-SRAM is proposed for the 0.5V operation low power CPU and SoC. During the read/hold, the V_{TH} of Fe-FETs automatically changes to increase the static noise margin by 60%. During the stand-by, the V_{TH} increases to decrease the leakage current by 42%. As a result, the supply voltage decreases by 0.11V, which decreases the active power by 32%.

INTRODUCTION

In the last five years, as the data through internet increases, the power consumption at the data center doubled. To solve the power crisis SSD is expected to replace HDD. For such an enterprise SSD, the Fe-NAND flash memory [1] is most suitable due to a low power consumption and a high reliability. As shown in Fig. 1(a), the Fe-NAND is composed of series-connected MFIS (Metal Ferroelectric Insulator Semiconductor) transistors (Fe-FETs). In the Fe-FET, the ferroelectric layer, $SrBi_2Ta_2O_9$ (SBT), is integrated in a gate stack of standard CMOS transistors with the metal gate, Pt and the high-K dielectric, Hf-Al-O (Figs. 1(b) and (c)). The program/erase voltage decreases from 20V of a conventional floating-gate (FG-) NAND to 6V. In the Fe-NAND, the electric polarization in the ferroelectric layer flips with a lower electric field and the threshold voltage, V_{TH} of a memory cell shifts (Fig. 1(d)). Due to a low program/erase voltage, a low power operation is achieved. Table 1 compares the conventional FG-NAND with the Fe-NAND. In the Fe-NAND, a high write/erase endurance, 100Million cycles is realized (Fig. 1(e)) because there is no stress-induced leakage current which degrades the data retention of the FG-NAND. An excellent 10year data retention is achieved by inserting a Hf-Al-O buffer layer between a ferroelectric layer and a Si substrate (Fig. 1(f)) [2]. The Fe-NAND is in principle scalable below 10nm to the crystal unit-cell size as a data is stored with an electric polarization in a ferroelectric gate insulator.

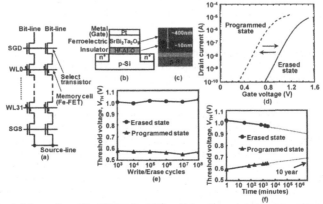

Fig. 1 Schematics of Fe-NAND/Fe-FET and the characteristics of Fe-FET [1].

Table 1 Comparison of conventional FG-NAND and Fe-NAND characteristics.

	Conventional Floating-Gate NAND	Proposed Fe-NAND
Write/Erase voltage	20V	6V
Endurance	10K (MLC) 100K (SLC)	100Million
Data retention	10 years	10 years

In this paper, the Fe-NAND flash memory with a non-volatile (NV) page buffer [3], a zero V_{TH} Fe-NAND memory cell scheme [4], a 1.2V operation adaptive charge pump circuit with Fe-FET diodes [5] and ferroelectric 6T-SRAM [6] are overviewed as recent studies of Fe-FETs. The technologies introduced in this paper aims for Fe-NAND flash memory for low power enterprise SSDs and a ferroelectric 6T-SRAM for 0.5V operation low-power CPU and SoC.

Fe-NAND FLASH MEMORY WITH NON-VOLATILE PAGE BUFFER

The critical problem of SSD is a slow random write. The write unit in a NAND flash memory is a page, 4-8KBytes. A page is composed of memory cells sharing a word-line and written only once to avoid a program disturb. For a PC and a data center application, the minimum write unit of OS is a sector, 512Bytes. 50% of data written by OS are less than 8sectors, 4KBytes. A random write to write a smaller data than a page size frequently happens. In case OS writes one sector in SSD, the remaining 80% of the page becomes a garbage (Fig. 2). As a garbage cumulates, a garbage collection is performed to increase a workable memory capacity. The garbage collection takes as much as 100ms [7], which is 100 times longer than a page program time, 800us, and thus causes a serious performance degradation.

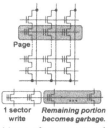

Fig. 2 Problems of one sector write in SSD.

Batch Write Algorithm

To solve the random write issue, a batch write algorithm is proposed (Fig. 3) [3]. A page buffer in the Fe-NAND temporarily stores a program data. To avoid a random write, the memory cell program starts after program data as much as the page size accumulate in page buffers. In *Step1* of Fig. 3, when OS issues one sector (512Bytes) write command to SSD, data are stored in page buffers but NOT programmed to memory cells. The NAND controller reports to OS that the write completed although the memory cell program is delayed to *Step4*. In *Step2*, when OS issues the second sector write command, the second data are also temporarily stored in page buffers. After data with a page size input to page buffers in *Step3*, the memory cell program starts in *Step4*. The proposed batch write algorithm eliminates the data fragmentation of a page during the random write. Considering the sequential write of SSD is over ten times faster than the random write, the batch write algorithm can at least doubles the SSD speed.

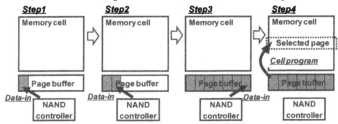

Fig. 3 Proposed batch write algorithm.

Non-volatile (NV) Page Buffer

The problem of the batch write algorithm is the data loss in a power outage. If a power outage happens in *Step4* of Fig. 3, all data in page buffers are lost since page buffers are volatile latch circuits. The computer system fails because OS recognizes that the program in *Step1-3* completed successfully. To solve this problem, a non-volatile (NV) page buffer is proposed. The NV-page buffer consists of a volatile *Latch* and a *NV-latch* (Fig. 4(a)). The *NV-latch* is realized with no additional process and the area penalty for the *NV-latch* is less than 1%. In the normal operation, the memory cell is programmed based on data in the *Latch* similarly as the conventional NAND. When the power outage occurs, data in the *Latch* is transferred to the *NV-latch* to avoid a data loss (Fig.4 (b)). In the next power-on, the data in the *NV-latch* is restored to the *Latch* (Fig.4 (c)). Then, the memory cell program is performed based on the data in the *Latch*.

The data storing operations are performed by using a charge remained in the chip. Although a charge is consumed at the charge pump to supply erase/program voltage, the internal supply voltage (V_{DD}) drop is 0.1V with conventional on-chip V_{DD} capacitors.

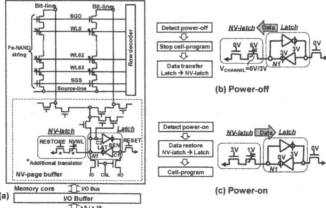

Fig. 4 Proposed NV-latch for page buffer.

Measurement Results

Fig. 5(a) shows measured program/erase characteristics of the *NV-latch*. A 10μs pulse shifts the V_{TH} of the *NV-latch*. Fig. 5(b) shows the measured program disturb. The measured V_{TH} shift caused by the program disturb is 0.07V and the program inhibit is realized.

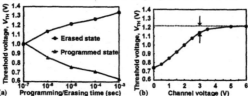

Fig. 5 Measured NV-latch program/erase characteristics and program disturb characteristics.

ZERO V_{TH} MEMORY CELL Fe-NAND FLASH MEMORY

In the Fe-NAND, the lower program voltage enables a lower power consumption but the reduced margin between the program voltage and the read voltage/program inhibit voltage degrades the read/program disturb. The data retention, read disturb and program disturb of the positive, zero and negative V_{TH} cells for the Fe-NAND flash memory (Fig. 6) are experimentally studied to clarify the best V_{TH} settings for Fe-NAND [4].

136

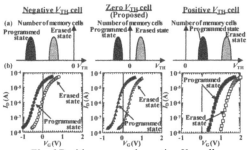

Fig. 6 Positive, zero and negative V_{TH} cell.

Data Retention Characteristics

It is known that the depolarization field, E_{DEP} [8] makes worse the data retention since it is in the opposite direction against an electric polarization existing in the ferroelectric layer. The magnitude of E_{DEP} depends on the V_{TH} due to the work function difference.

Fig. 7(a) shows the measured data retention where V_G is fixed at 0V. Fig. 7(b) illustrates the polarization-electric field curve. Negative V_{TH} cell is expected to show large E_{DEP} at erased state which results in short erased state retention. Similarly, positive V_{TH} cell is expected to show large E_{DEP} at programmed state. Fig. 7(c) shows the V_{TH} shift (ΔV_{TH}) after 10-year data retention extracted from Fig. 7(a) for each programmed and erased state. Table 2 summarizes the measured total V_{TH} shift, ΔV_{TH_ALL}, sum of ΔV_{TH} for programmed and erased states after a 10-year data retention. The proposed zero V_{TH} cell shows the smallest ΔV_{TH_ALL} of 0.23V that is 10% smaller than the conventional positive V_{TH} cell. In the negative V_{TH} cell, ΔV_{TH_ALL} is 0.49V and unacceptably large due to the enlarged E_{DEP} for the erased state.

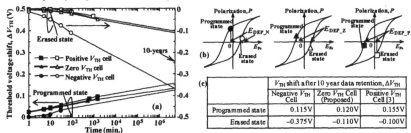

	V_{TH} shift after 10 year data retention, ΔV_{TH}		
	Negative V_{TH} Cell	Zero V_{TH} Cell (Proposed)	Positive V_{TH} Cell [3]
Programmed state	0.115V	0.120V	0.155V
Erased state	−0.375V	−0.110V	−0.100V

Fig. 7 Retention characteristics of Fe-FET for negative/zero/positive V_{TH} cell.

Disturb Characteristics

Fig. 8(a) shows the read disturb. During the read, unselected word-lines are biased to V_{READ} which is higher than the memory cell V_{TH} so that unselected cells turn on. As shown in Fig. 8(b), the unselected cells are in a weak program condition and the V_{TH} of the erased cells decreases (Fig. 8(c)). Fig. 8(d) shows the measured V_{TH} shift, ΔV_{TH}, due to the read disturb. The read stress time is 10^4 sec that corresponds to the 10^9 read cycles with a 10μs read pulse. The 10^9 read cycles are sufficient for the enterprise SSD application. For the positive, zero and negative

V_{TH} cells, V_{READ} are 1.7V, 0.7V and 0.4V, respectively. Measured ΔV_{TH} are 0.205V, 0.14V and 0.31V for the positive, zero and negative V_{TH} cells, respectively. As the V_{TH} of the memory cell decreases, V_{READ} becomes lower. As a result, the read disturb decreases. In the zero V_{TH} cell, ΔV_{TH} is suppressed to 0.14V which is 32% smaller than that of the conventional positive V_{TH} cell.

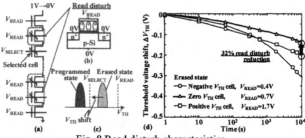

Fig. 8 Read disturb characteristics.

The program disturb of a Fe-NAND cell is shown in Fig. 9(a)-(c). The selected word-line is biased to the program voltage, V_{PGM}, 6V. The selected and unselected bit-lines are biased to 0V and 3V. Unselected word-lines are applied to V_{PASS}. In the selected *Cell A*, the electric polarization in the ferroelectric layer flips. *Cell B* and *Cell C* suffers from the V_{PASS} disturb and V_{PGM} disturb, respectively.

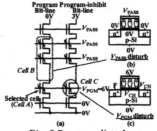

Fig. 9 Program disturb.

As shown in Fig. 10(a), the measured ΔV_{TH} of the V_{PASS} disturb is almost the same for the three V_{TH} cells. Fig. 10(b) shows the measured program disturb. To suppress the V_{PGM} disturb, V_{PASS} must be large enough to transfer 3V to the channel of *Cell C*. The intersection of the V_{PGM} disturb and the V_{PASS} disturb is where the program disturb is minimized. As the V_{TH} of the memory cell decreases, V_{CH} becomes higher and the V_{PGM} disturb decreases. In the proposed zero V_{TH} cell, the measured ΔV_{TH} is suppressed to 0.13V, which is 24% smaller than that of the conventional positive V_{TH} cell. Table 2 summarizes the work in this section. Zero V_{TH} cell shows the least ΔV_{TH} in retention and read disturb characteristics while showing moderate program disturb characteristics.

138

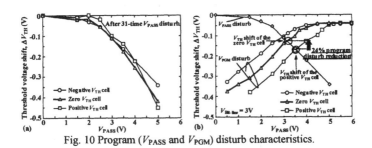

Fig. 10 Program (V_{PASS} and V_{PGM}) disturb characteristics.

Table 2 Summary of the retention and disturb characteristics of the three V_{TH} cells.

	Total V_{TH} shift, ΔV_{TH_ALL}		
	Negative V_{TH} Cell	Zero V_{TH} Cell (Proposed)	Positive V_{TH} Cell [3]
Data retention	0.490V (192%)	0.230V (90%)	0.255V (100%)
Read disturb	0.310V (151%)	0.140V (68%)	0.205V (100%)
Program disturb	0.110V (65%)	0.130V (76%)	0.170V (100%)

1.2V OPERATION HIGH EFFICIENCY ADAPTIVE CHARGE PUMP CIRCUIT

Highly efficient charge pump circuit is essential in NAND flash memories to generate high voltage for read/program operation. To realize the low voltage and low power Fe-NAND flash memories, an adaptive charge pump circuit is proposed [5]. The proposed charge pump increases the power efficiency and the output voltage by 143 and 25% without the circuit area and the process step penalty.

Conventional Charge Pump Circuits

In the 1.2 V power supply Fe-NAND flash memory, the V_{READ} and the V_{PGM} charge pump circuits generate the read voltage, 2 V and the program voltage, 6 V from V_{DD}, 1.2 V, respectively. The V_{READ} and the V_{PGM} charge pump circuits are composed of the two-stage and the eight-stage charge pump stages, respectively. Figs. 11(a) and (b) show the conventional charge pump circuits. The V_{TH} of MOS diodes at all pump stages are fixed at V_{TH_MOS}. In the conventional charge pump circuits, since the V_{TH} of MOS diodes are fixed at a predetermined value, it was impossible to realize a high output voltage or a high power efficiency for both the V_{READ} and the V_{PGM} charge pumps at the same time. As shown in Figs. 12(a) and (b), V_{TH_MOS} has optimal values to maximize the power efficiency or the output voltage, V_{OUT}. Further, the optimal V_{TH} is also different between V_{PGM} and V_{READ} charge pump. This is because the body effect is stronger at the latter stage of V_{PGM} pump than the V_{READ} pump. If V_{TH_MOS} is -0.3 V to maximize the power efficiency of the V_{READ} charge pump, the power efficiency of the V_{PGM} charge pump is 127% smaller than the optimal value with V_{TH_MOS}=-0.6 V. Similarly, if V_{TH_MOS} is -0.5 V to maximize V_{OUT} of the V_{READ} charge pump, V_{OUT} of the V_{PGM} charge pump decreases by 19% from its maximum value with V_{TH_MOS}=-1.0 V.

Proposed Dual V_{TH} Charge Pump Circuits

To increase the power efficiency and V_{OUT} of the V_{PGM} charge pump, this paper proposes a dual V_{TH} charge pump circuit as shown in Figs. 11(c) and (d). Here, Fe-FETs are used as diodes in the charge pump. There is no area or process overhead as Fe-FETs in the proposed charge pump are fabricated with the same process steps as the Fe-NAND memory cells. The V_{TH} of Fe-FETs is adjusted to maximize the power efficiency or V_{OUT}. The V_{TH} adjustment is performed after the chip fabrication using the testing equipment. The V_{TH} of the V_{READ} and the V_{PGM} charge pumps are represented as V_{TH_READ} and V_{TH_PGM}. As shown in Figs. 12(a) and (c), by changing the V_{TH} of Fe-FETs, V_{TH_READ} and V_{TH_PGM} are independently optimized at -0.3 and -0.6 V. As a result, the power efficiency of the V_{PGM} charge pump increases by 127%. By selecting V_{TH_READ} and V_{TH_PGM} as -0.5 and -1.0 V, V_{OUT} of both the V_{READ} and the V_{PGM} charge pumps are also maximized. V_{OUT} of the V_{PGM} charge pump increases by 19%.

Proposed Adaptive Charge Pump Circuits

To further increase the power efficiency and V_{OUT}, this paper also proposes an adaptive charge pump as shown in Figs. 11(e) and (f). The V_{TH} of both the V_{READ} and the V_{PGM} charge pumps are different with V_{TH_Fe1}, V_{TH_Fe2}, ... V_{TH_Fe8} at each pump stage. V_{TH_Fe1}, V_{TH_Fe2}, ... V_{TH_Fe8} has a relationship of $V_{TH\_Fe1} > V_{TH\_Fe2} > ... > V_{TH\_Fe8}$ because of the body effect. Fig. 13 shows the power efficiency and V_{OUT} of the adaptive V_{PGM} charge pump as functions of V_{TH_Fe1} and V_{TH_Fe2}. When V_{TH_Fe1} and V_{TH_Fe2} are optimized, the power efficiency and V_{OUT} are simulated with the conditions of the fixed V_{TH_Fe3}, V_{TH_Fe4}, ... V_{TH_Fe8}. The dotted line shows the point where V_{TH_Fe1} equals to V_{TH_Fe2}. The V_{TH} optimal point of the adaptive charge pump is out of the dotted line. By using different values of V_{TH_Fe1} and V_{TH_Fe2}, the adaptive charge pump can optimize V_{TH} more efficiently than the dual V_{TH} charge pump. Compared with the dual V_{TH} charge pump, the adaptive charge pump increases the power efficiency and V_{OUT} by 6.9 and 4.5%, respectively.

Table 3 summarizes the comparison of the conventional eight-stage charge pump and the proposed charge pumps. The proposed adaptive charge pump is most power efficient and has the highest V_{OUT} as the V_{TH} at each pump stage is best optimized.

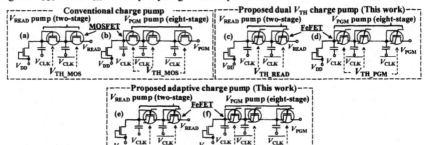

Fig. 11 Conventional, proposed dual V_{TH} and proposed adaptive charge pump circuits.

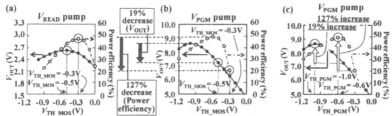

Fig. 12 Simulated output voltage, V_{OUT}, and power efficiency. V_{DD}=1.2V. f_{CLK}=3MHz. (a), (b) Conventional V_{READ} and V_{PGM} pump, (c) Proposed dual V_{TH} pump.

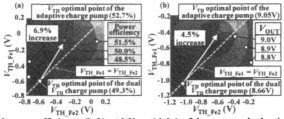

Fig. 13 Simulated power efficiency (left) and V_{OUT} (right) of the proposed adaptive charge pump.

Table 3 Comparison of the conventional and the proposed charge pump circuits.

	Conventional charge pump	Proposed dual V_{TH} charge pump	Proposed adaptive charge pump
Power efficiency (%)	21.7 (x1.00)	49.3 (x2.27)	52.7 (x2.43)
V_{OUT} (V)	7.26 (x1.00)	8.66 (x1.19)	9.05 (x1.25)
Circuit area (a.u.)	1	1	1

Measurement Results

The proposed two-stage charge pump circuit is fabricated with the CMOS compatible Fe-FET process. Fig. 14(a) shows the measured V_{OUT} as a function of the V_{TH} in the one-stage adaptive charge pump. As the V_{TH} decreases, V_{OUT} increases. Fig. 14(b) shows that there is no V_{TH} shift of Fe-FETs due to the disturb during the boosting. Figure 14(c) describes the bias condition of Fe-FET. While a voltage higher than V_{DD} is biased to Fe-FET, the V_{TH} shift of Fe-FET is negligibly small irrespective of the operation time and V_{DD} and a highly reliable operation is realized.

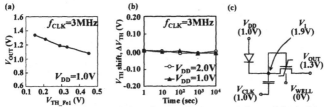

Fig. 14 Measured (a) pump characteristics, (b) disturb characteristics and (c) its bias condition.

0.5V OPERATION Fe-FET 6T-SRAM WITH V_{TH} SELF-ADJUSTING FUNCTION

Decreasing the supply voltage (V_{DD}) is essential to reduce the power consumption of CPU and SoC. To realize a low voltage/power CPU and SoC, SRAM with a very low supply voltage e.g. 0.5V capability is required. The stability of the SRAM is characterized by static noise margin (SNM). SNM represents the diagonal length of the largest square in the butterfly curves of two inverters of the SRAM cell. When the SNM is 0V, the read operation fails. SNM degrades by lowering V_{DD}. It also degrades by the presence of V_{TH} variations of the MOSFETs in SRAM cells. As the feature size of SRAM decreases to sub-30nm, the V_{TH} variation of the MOSFETs due to the random dopant fluctuation significantly increases and the SNM severely degrades. As a result, decreasing V_{DD} of SRAM is difficult in the nano-scale CMOS circuits and V_{DD} of CPU and SoC cannot be decreased. To overcome this problem, ferroelectric 6T-SRAM is proposed [6].

Proposed Ferroelectric 6T-SRAM

The proposed 6T-SRAM is shown in Fig. 15. The proposed SRAM consists of 6 transistors, PU1, 2, PD1, 2, PG1, 2, which is identical to the conventional 6T-SRAM. The body of NMOS and PMOS are set to V_{DD} and V_{SS} to self-adjust the V_{TH} of PU1, 2 and PD1, 2 and increase SNM. The forward-biased diode current in the proposed SRAM is suppressed at 0.5V V_{DD}. The V_{TH} of Fe-FETs changes by controlling the electric field between the gate and the body/channel. In the NMOS, PD1, 2 when the gate-body voltage (V_{GB}) of -0.5V is applied, the negative polarization is located close to the body and the V_{TH} increases. When the gate-channel voltage (V_{GC}) of 0.5V is applied to the NMOS, the positive polarization is located close to the body and the V_{TH} decreases. Similarly, the $|V_{TH}|$ of PMOS, PU1, 2 increases when V_{GB} of 0.5V is applied and decreases when V_{GC} of -0.5V is applied.

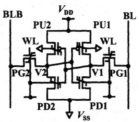

Fig. 15 Proposed ferroelectric 6T-SRAM.

Static Noise Margin Enhancement by V_{TH} Self-Adjustment

Fig. 16 describes the SNM increase. When the SRAM cell holds the "0"-data (Fig. 16(a), (c)) in the right inverter, the $|V_{TH}|$ of PU1 increases by the gate-body electric field of 0.5V V_{GB} and the V_{TH} of PD1 decreases by the gate-channel electric field of 0.5V V_{GC}. The body voltage (0.5V) of PD1 is shielded by the grounded channel and thus does not affect the V_{TH} shift. In the left inverter, the $|V_{TH}|$ of PU2 decreases by the gate-channel electric field of -0.5V V_{GC} and the V_{TH} of PD2 increases by the gate-body electric field of -0.5V V_{GB}. The body voltage (0V) of PU2 is shielded by the channel with 0.5V. As a result, the trip point of the right inverter decreases and the trip point of the left inverter increases. The storage nodes, V1 and V2 become more likely to hold 0V and 0.5V, respectively. As a result, holding "0"-data becomes more stable, resulting in the increase of SNM. In case of "1"-data (Fig. 16(b), (d)), the V_{TH} shifts in an opposite direction

142

as "0"-data, increasing SNM for "1"-data. The Monte Carlo simulation results with 100 time iteration and 30 mV V_{TH} variation are shown in Figs. 17(a) and (b). In the proposed cell, SNM with the V_{TH} variation increases by 546% from 13 mV to 84 mV. SNM without the V_{TH} variation increases by 60% (Fig. 17(c)).

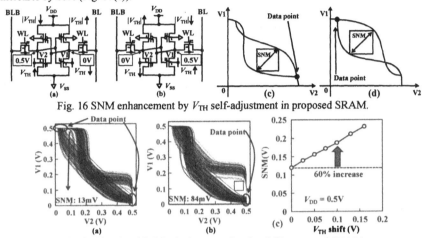

Fig. 16 SNM enhancement by V_{TH} self-adjustment in proposed SRAM.

Fig. 17 Simulated SNM of (a) conventional and (b) proposed SRAM. (c) SNM enhancement as a function of V_{TH} shift.

Measured hold SNM and the bias condition are shown in Fig. 18. Due to the automatic V_{TH} shift, the measured SNM increases from 1.38 V to 1.62 V. By increasing the electric field across the ferroelectric layer, the automatic V_{TH} shift could be even enlarged.

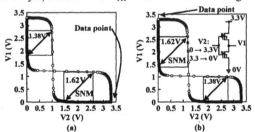

Fig. 18 Measured hold SNM in ferroelectric inverter.

Leakage Current Reduction

The subthreshold current of the off-state FETs shown in Fig. 19(a) determines the leakage current of the SRAM cell. During the stand-by, the leakage PMOS and NMOS are automatically set to the high $|V_{TH}|$ state to reduce the leakage current. Fig. 19(b) shows the measured leakage current reduction. During the stand-by, as the V_{TH} of the leakage transistors increases from 0.2 V to 0.3 V, the leakage current of the proposed SRAM decreases by 42%.

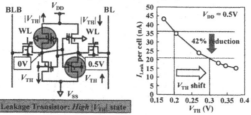

Fig. 19 Leakage reduction in proposed SRAM.

V_{DD} Scaling and Active Power Reduction

Minimum operation voltage V_{DDmin} of the proposed SRAM is shown in Fig. 20. The read/write margin, leakage and read speed restrictions are taken into account. As shown in Fig. 20(a), an excessive V_{TH} shift causes a write failure and increases the minimum operating voltage V_{DDmin}. At 0.5V V_{DD} operation SRAM, the optimal V_{TH} shift is 0.1V and V_{DD} of the proposed cell decreases by 0.11V. The V_{DD} scaling from 0.61V to 0.5V reduces the active power consumption, $f \times C \times V_{DD}^2$, by 32% (Fig. 20(b)).

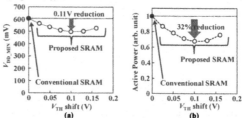

Fig. 20 V_{DDmin} reduction and active power reduction in proposed SRAM.

CONCLUSIONS

The recent studies of Fe-FET is overviewed. The Fe-NAND flash memory with a non-volatile (NV) page buffer, a zero V_{TH} Fe-NAND memory cell scheme, a 1.2V operation adaptive charge pump circuit with Fe-FET diodes and ferroelectric 6T-SRAM are introduced. Fe-FET is suitable for realizing low power enterprise SSDs and low-voltage CPU and SoC applications.

ACKNOWLEDGMENTS

This work was partially supported by NEDO.

REFERENCES

[1] S. Sakai et. al., *NVSMW&ICMTD*, pp.103-105, 2008, [2] S. Sakai et al., *Electron Device Lett.*, vol. 25, pp.369-371, 2004. [3] T. Hatanaka et. al., *Symp. VLSI Circuits*, pp.78-79, 2009. [4] T. Hatanaka et. al., *ESSDERC*, pp.225-228, 2009. [5] S. Noda et. al., *SSDM*, pp.162-163, 2009. [6] S. Tanakamaru et. al., *IEDM*, pp.283-286, 2009. [7] K.Takeuchi, et. al., *ISSCC*, pp.144-145, 2006. [8] T. P. Ma et. al., *Electron Device Lett.*, vol. 23, pp.386-388, 2002.

Mater. Res. Soc. Symp. Proc. Vol. 1250 © 2010 Materials Research Society　　　　　　1250-G13-07

Fabrication of IGZO and In$_2$O$_3$-Channel Ferroelectric-Gate Thin Film Transistors

Eisuke Tokumitsu and Tomohiro Oiwa
Precision and Intelligence Laboratory, Tokyo Institute of Technology
4259-R2-19 Nagatsuta, Midori-ku Yokohama 226-8503, Japan

ABSTRACT

We have fabricated and characterized ferroelectric-gate TFTs using In-Ga-Zn-O (IGZO) or In$_2$O$_3$ as a channel material. The ferroelectric gate insulator used in this work is (Bi,La)$_4$Ti$_3$O$_{12}$ (BLT). We observed normal n-channel transistor operation for both IGZO and In$_2$O$_3$-channel TFTs. However, a charge injection type hysteresis was observed for IGZO channel TFTs in drain current – gate voltage (I$_D$-V$_G$) characteristics. Post fabrication anneal at 300°C reduced the charge-injection-tyoe hystereesis and the subthreshold swing was also improved from 0.27 to 0.19 V/decade. On the other hand, when the In$_2$O$_3$ was used as a channel, hysteresis due to the ferroelectric gate insulator was clearly observed in I$_D$-V$_G$ characteristics. A memory window of 2V, a subthreshold voltage swing of 0.35V/decade, a field-effect mobility of 1.6 cm^2/Vs, and an on/off drain current ratio of more than 10^6 were obtained.

INTRODUCTION

Oxide-channel thin film transistors (TFTs) have attracted much attention for flat-paneel display applications. In particular, ZnO[1,2] and amorphous In-Ga-Zn-O (IGZO)[3,4] are the most mromising candidates as channel materials, because TFTs with such oxides exbihit a channel mobility of more than 10 cm^2/Vs, which is much higher than amorphous silicon TFTs. Hence, it will be possible to implement driver and peripheral logic circuits by IGZO-based TFTs. In addition, it will be interesting if we can realize high-density nonvolatile memories as well as logic circuits and driver TFTs by all-oxide TFTs on a single substrate for future system-on-panel or system-on film applications. In our previous work, we reported excellent electrical properties of indium-tin-oxide (ITO) channel ferroelectri-gate TFTs using (Bi,La)$_4$Ti$_3$O$_{12}$ (BLT) gate insulator, where we demonstrated that the ferroelectric gate insulator can control large channel charges of the conductive ITO film[5,6]. In this work, we have fabricated and characterized ferroelectric-gate TFTs using IGZO or In$_2$O$_3$ as a channel material, because lower carrier concentration can be expected for these films than ITO.

EXPERIMENTAL PROCEDURE

Bottom-gate structure TFTs were fabricated on SiO$_2$/Si substrates using IGZO or In$_2$O$_3$ as a channel and BLT as a gate insulator. First, Pt/Ti or ITO layer was deposited by sputtering and patterned to form the bottom gate electrodes. Next, the ferroelectric BLT thin film was formed by the sol-gel technique. Bi$_{3.35}$La$_{0.75}$Ti$_3$O$_{12}$ source solution was spin-coated and consolidated at 400 °C for 10 min in O$_2$, and then crystallized at 750°C for 30 min in O$_2$ ambient.

The thickness of the BLT film is approximately 200 nm. Then, thin (10-20 nm) IGZO or In_2O_3 channel layer was deposited by sputtering at room temperature. The carrier concentration of the sputtered IGZO film is sensitive to the O_2/Ar ratio during the deposition, which were adjusted to obtaine a carrier concentration of about 10^{18} cm^{-3}. Next, source and drain electrodes were deposited and the device was isolated by the dry etching technique. The samples were annealed up to 300°C to improve the electrical characteristics. The channel length and width are typically 10 and 100 μm, respectively.

RESULTS AND DISCUSSION

We first examin the deposition conditions of IGZO film by sputtering. Figure 1 shows resisitivity of the IGZO films measured by four-probe method as a function of $O_2/(Ar+O_2)$ ratio during the deposition. IGZO film was deposited at room temperature and the RF power and total pressure during the deposition are 300W and 0.5 Pa, respectively. It is shown that the resisitivity of the IGZO film drastically changes by about 5 orders of magnitude, when the $O_2/(Ar+O_2)$ ratio is increased from 0 to 3%. This results is sumilar to the previous report [7]. It is also found form the Hall measurements that the carrier concentration and Hall mobility of the IGZO film fabricated at a $O_2/(Ar+O_2)$ ratio of 1.9% are $2x10^{18}$ cm^{-3} and 9 cm^2/Vs, respectively, whereas those of IGZO film fabricated without introducing oxygen are $1.2x10^{19}$ cm^{-3} and 15 cm^2/Vs, respectivly. Hence, we used the deposition condition of $O_2/(Ar+O_2)$ ratio of 1.9% for TFT fabrication.

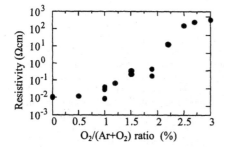

Figure 1. Resisitivity of the IGZO film fabricated by sputtering as a function of $O_2/(Ar+O_2)$.

Figure 2 shows polarization – electric field (P-E) hysteresis loops of the ferroelectric BLT film used for the TFT fabrication. The film was deposited on $Pt/Ti/SiO_2/Si$ substrate by the sol-gel technioque at a crystallization temperature of 750°C. A remanent polarization of 15 $\mu C/cm^2$ was obtained, which is a typical value for BLT films.

Next, we fabricated IGZO/BLT structure TFTs using the IGZO and BLT films shown above. IGZO channel layer was deposited with a $O_2/(Ar+O_2)$ ratio of 1.9% and a RF power of 300 W, which resulted in a carrier concentration of $2x10^{18}cm^{-3}$. Figures 3(a) and 3(b) show drain current – gate voltage (I_D-V_G) and drain current – drain voltage (I_D-V_D) characteristics of the IGZO/BLT TFT where the IGZO channel layer was deposited at room temperature. We observed

146

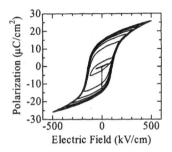

Figure 2. P-E hysteresis loops of BLT film used for the TFT fabrication.

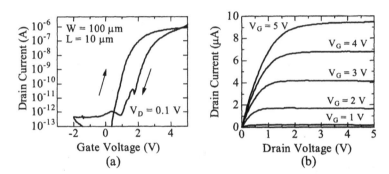

(a)   (b)

Figure 3. (a) I_D–V_G and (b) I_D–V_D characteristics of the IGZO/BLT TFTs where IGZO channel layer was deposited at room temperature.

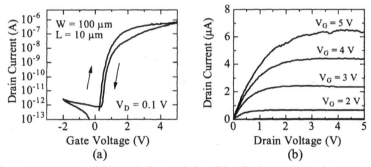

(a)   (b)

Figure 4. (a) I_D–V_G and (b) I_D–V_D characteristics of the IGZO/BLT TFTs after TFT was annealed at 300°C.

normal n-channel transistor operation for both IGZO/BLT TFTs. On/off current ratio is more than 10^6. However, when the IGZO film was deposited at room temperature, hysteresis due to charge injection was observed in I_D-V_G characteristics in spite of the ferroelectric gate insulator. This suggests the presence of interfacial traps between the BLT and IGZO and also some defects may exisit in the IGZO film itself.

To improve the electrical properties, we carried out post fabrication annealing. Figures 4(a) and (b) show I_D-V_G and I_D-V_D characteristics of the IGZO/BLT TFT which was annealed at 300°C for 15 min in O_2 ambient after the device fabrication. It is shown that charge injection-type hysteresis in I_D-V_G curve was surpressed and that the subthreshold voltage swing was improved from 0.27 to 0.19 V/decade. However, non-volatile memory effect due to the ferroelectric BLT gate insulator still was not observed. To investigate the damage or reaction which may be introduced during the IGZO deposition by sputtering and post annealing, we observed the cross section of IGZO/BLT structure after 300°C annealing by transmission electron microscope (TEM) as shown in figure 5. It is found that about 20 nm-thick IGZO was conformally deposited on the polycrystalline BLT film and that there is a bright interfacial layer between IGZO and BLT. Energy dispersive X-ray spectrometry (EDX) analysis revieled that the Zn was segregated at the interface. The composition at the point A near the surface was measured as In:Ga:Zn:O = 15:13:8:64, whereas that at the point B near the interface was In:Ga:Zn:O = 8:9:24:59. This segregation is not so serious when the IGZO film was deposited on SiO_2/Si substrate. Hence, there was some interaction between IGZO and BLT after the structure was annealed at 300°C.

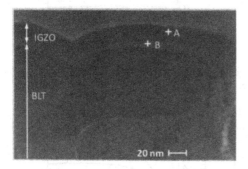

Figure 5. TEM cross section of IGZO/BLT interface.

Since we observed a siginificant Zn segregation at the IGZO/BLT interface, we next examined In_2O_3 as a channel layer which does not contain Zn. In_2O_3 layer was deposited at room temperature with an RF power of 50 W and an $O_2/(Ar+O_2)$ ratio of 3 %, which results in a carrier concentration of around 10^{18} cm^{-3}. Figures 6(a) and (b) show I_D-V_G and I_D-V_D characteristics of the In_2O_3/BLT TFT which was annealed at 300°C for 15 min in O_2 ambient after the device fabrication. A clear counter-clockwise hysteresis loop was observed in I_D-V_G characteristics due to the ferroelectric gate insulator, which confirm the nonvolatile memory function of the device. In addition, a large on/off drain current ratio of more than 10^6, a memory

148

window of 2V, a subthreshold voltage swing of 0.35 V/decade were obtained. The field effect mobility was estimated from 1.3 to 1.6 cm^2/Vs.

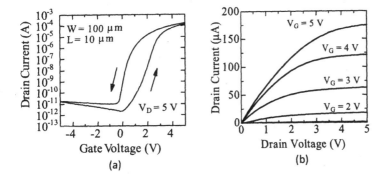

Figure 6. (a) I_D–V_G and (b) I_D–V_D characteristics of the In_2O_3/BLT TFTs after TFT was annealed at 300°C.

CONCLUSIONS

Bottom-gate-stucure TFTs using a ferroelectric BLT gate insulator and IGZO or In_2O_3 channel were fabricated and characterized. Ferroelectric BLT films were prepared by the sol-gel technique, whereas IGZO or In_2O_3 channel layer was deposited by sputtering. Normal n-channel transistor operations were observed for both IGZO and In_2O_3-channel TFTs. When the IGZO film was used as a channel layer, a charge injection type hysteresis was observed in I_D-V_G characteristics even after post fabrication anneal at 300°C and non-volatile memory effect due to the ferroelectric BLT gate insulator was not observed. On the other hand, for In_2O_3-channel TFTs, a hysteresis loop due to the ferroelectric gate insulator was clearly observed in I_D-V_G characteristics. A memory window of 2V, a subthreshold voltage swing of 0.35V/decade, a field-effect mobility of 1-2 cm^2/Vs, and a on/off drain current ratio of more than 10^6 were obtained.

ACKNOWLEDGMENT

This work was partially supported by Grant-in-Aid for Scientific Research (B) (21360144) from Japan Society for the Promotion of Science.

REFERENCES

1. P. F. Carcia, R. S. McLean, M. H. Reilly, and G. Nunes, Jr., Appl. Phys.Lett. **82**, 1117 (2004)
2. J. Nishii, F. M. Hossain, S. Takagi, T. Aita, K. Saikusa, Y. Ohmaki, I.Ohkubo, S. Kishimoto, A. Ohtomo, T. Fukumura, F. Matsukura, Y. Ohno, H. Koinuma, H. Ohno, and M. Kawasaki,

Jpn. J. Appl. Phys., **42**, L347 (2003).
3. K. Nakajima, H. Ohta, K. Ueda, T. Kamiya, M. Hirano, and H. Hosono, Science **300**, 1269 (2003).
4. K. Nomura, H. Ohta, A. Takagi, T. Kamiya, M. Hirano, and H. Hosono, Nature **432**, 488 (2004).
5. T. Miyasako, M. Senoo, and E. Tokumitsu, Appl. Phys. Lett., **86**, 162902 (2005).
6. E. Tokumitsu, M. Senoo and T. Miyasako, J. of Microelectronic Engineering, **80**, 305, (2005).
7. H. Yabuta, M. Sano, K. Abc, T. Aiba, T. Den, H. Kumomi, K. Nomura, T. Kamiya, and H. Hosono, Appl. Phys. Lett., **89**, 112123 (2006).

Organic Memory

Mater. Res. Soc. Symp. Proc. Vol. 1250 © 2010 Materials Research Society   1250-G04-07

Study on C60 doped PMMA for organic memory devices

Micaël Charbonneau[1], Raluca Tiron[1], Julien Buckley[1], Mathieu Py[1], Jean Paul Barnes[1], Samir Derrough[1], Christophe Constancias[1], Christophe Licitra[1], Claire Sourd[1], Gérard Ghibaudo[2], Barbara De Salvo[1]
[1]CEA, LETI, MINATEC, 17 rue des Martyrs, 38054 Grenoble cedex 9, France
[2]IMEP-LAHC, CNRS, MINATEC, 38016 Grenoble, France

ABSTRACT

In this paper, fabrication of nanocomposite thin films with introduction of fullerene (C60) molecules in PolyMethyl MethAcrylate resist (PMMA) was investigated from a material and electrical point of view. The effective inclusion of C60 molecules in the samples was characterized by using UV-vis and Tof-SIMS instruments. The modified resist PMMA:C60 was also studied with Thermal analysis (TGA, DSC) where modification of physical properties is reported. Films were included in MIS and MIM devices and results on non volatile trapping of C60 doped PMMA are presented. Moreover, e-beam exposure tests showed that PMMA:C60 active layers for memory devices, were scalable in size.

INTRODUCTION

The recent growth of interest in so called "organic electronics" opens new opportunities to investigate blends of well-known materials. In this context organic polymers and nanocomposites are very promising. Indeed research has already successfully demonstrated the use of organic bulk materials for the fabrication of electronic and optoelectronic devices [1,2]. Recently, several groups have also investigated polymeric materials for memory applications [3-4]. These materials offer important advantages in terms of fabrication methodology and properties tunability: wide versatility in chemical composition allowing large tailoring of electrical behavior, simple processable deposition methods, high flexibility and low cost. These materials could be included in different devices such as capacitors, transistors or resistors, suitable for different memory architectures [5]. In particular, MIS (Metal Insulator Semiconductor) [6] and MIM (Metal Insulator Metal) [7] structures have been largely studied in the literature, the insulator being implemented with macro-polymers, copolymers or nano-composites with metallic nano-particles or small organic molecule inclusions [5]. High attention has been given in the literature to compositions including C60 fullerenes in polymer thin films such as: PVK (poly(Nvinylcarbazole)), PVP (Poly-4-vinylphenol), PS (Polystyrene) and PMMA [5,6,7,8,9]. We propose in this work an extensive and novel study of PMMA-based nanocomposites with C60 inclusions (Figure 1-a).

EXPERIMENTS

Poly-Methyl-MethAcrylate polymer (PMMA) previously dissolved in Anisole (2 mg/mL; Mw = 950K) was purchased from MicroChem. The fullerene solution was obtained by dissolving C60 powder from Sigma Aldrich in Toluene (2.6 mg/mL). Then PMMA:C60 solutions were mixed from PMMA:Anisole, C60:Toluene and completed with pure Toluene to obtain the targeted amount of C60 and keep a constant Solvent\Materials ratio for the different

153

compositions. After sonication the resulting compositions ranging from r= 0wt% to 20wt% of C60 (where r = mC60/mPMMA) were spin-coated on primed 200mm Silicon wafers and dried at 120°C for 90s on a TEL MK8 Track. The thickness of the films was controlled by ellipsometry (Prometrix UV1280).

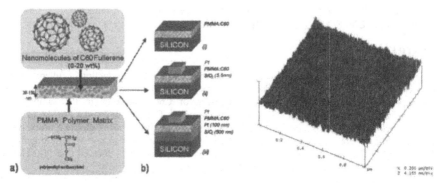

Figure 1: (a) Chemical Structure of PMMA and C60 fullerene. (b) Schema of: (i) structure suitable for physical/chemical characterization; (ii) MIS (Pt, PMMA/SiO2, p-Si) test structure; (iii) MIM (Pt, PMMA, Pt) test structure.

Figure 2: AFM image of PMMA:C60 r= 10wt%. The scan area is 1µmx1µm. The estimated Roughness is 0.29nm rms.

Different base wafers were used (figure 1-b): silicon substrate for physical and chemical characterization, 5.5 nm thermal oxide built on p-type silicon for MIS structures and full sheet 100 nm Pt bottom electrodes deposited by PVD on 500 nm thermal oxide coated silicon for MIM structures. To build the top electrode on MIM and MIS samples, square patterns (1x1 mm²) were performed on top of the polymeric film via shadow evaporation in a PLASYS 500s under 10^{-8} Torr. Pt metallization was used in order to avoid electrode oxidation during electrical characterization of the polymer.

Thin films on Silicon substrates (figure 1-b-i), were analyzed with UV visible ellipsometry and Tof-SIMS spectrometry. UV measurements were performed on a Woollam M2000 rotating compensator ellipsometer in the 190-1700 nm wavelength range. Spectra were recorded at three angles of incidence: 65°, 70° and 75°. Both refractive index and extinction coefficients were calculated in the ultraviolet range using mathematical inversion. Concerning ToF-SIMS, samples were profiled with an ION TOF ToF-SIMS V, using a 25 keV Bi3+ analysis beam and a 250 eV Cs+ sputtering beam both incident at 45° [10]. Thermal behavior of the samples was investigated through Differential Scanning Calorimetry (DSC 2920) and Thermo Gravimetric Analysis (TGA 2950) [11]. All instruments are from TA Instruments.

For e-beam exposure tests, resist patterns were written using VB6 ultra high resolution (UHR) LEICA Gaussian beam equipment with an acceleration voltage at 100 kV and a beam current of 1 nA. Spot measurements at 100 keV indicated a size of 5 nm using Si/SiGe dense grating method. Iso dens patterns (CD in the range 200nm -30nm) were exposed at 500µC/cm² - 700µC/cm² and then developed during 120s in 3:1 IPA:MIBK solvents.

C-V measurements of Metal Insulator Semiconductor and Metal Insulator Metal structures were performed with Agilent B1500A under nitrogen flux at 25°C.

RESULTS AND DISCUSSION

Physical and chemical characterizations

First, ellipsometry, profilometry, AFM and SEM were performed to control thin film fabrication. Smooth layers (< 0.4 nm rms) with targeted thickness ranging from 30 nm to 150 nm were obtained and revealed no significant roughness variation in samples with C60 content up to 20wt%. As example, an AFM image of PMMA with r=10wt% of C60 is reported on figure 2.

Since C60 doping in the PMMA insulating matrix should dictates the electrical behavior of the materials, physical and chemical characterizations were carried out to get information on thin film properties such as composition and C60 uniformity distribution. This is a challenging task as polymers and C60 are both carbon-based. First, effective inclusion of C60 in the PMMA film was analyzed by measuring UV/vis spectra. The Extinction coefficient k between 190 and 360 nm is reported on figure 3. The Inset shows the extinction coefficient related to the peak at λ=261nm. Considering a linear dependence, we show here that UV Visible spectroscopy is a good quantitative technique in order to quantify C60 loading rate in the film.

To go further, with ToF-SIMS measurements we successfully tagged signals related to C60 and PMMA Respectively C5H9O2- (m=101 u) for the PMMA monomer, and C60- (m= 720 u) for C60. The only difference detected between mass spectra of pure PMMA and blended layers consisted of this C60 related mass group, certifying their correlation with C60 presence in the blend but also showing the very little fragmentation of the fullerenes under abrasion and analysis beams. Extremely stable signals for organic related fragments in the estimated nanocomposite region indicates very little damaging of the organic layer while profiling (for more details see reference [10]). The extracted ratios of these signals are displayed on figure 4-a) and show a linear behavior thus demonstrating the ability to quantify precisely C60 weight ratio in the nanocomposite. Moreover this technique (based on gradual sample abrasion), lead to a depth profile of the sample. Here the constant C60 signals in the nanocomposite suggest a uniform depth distribution in the PMMA.

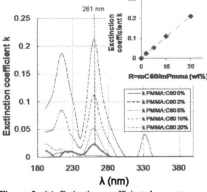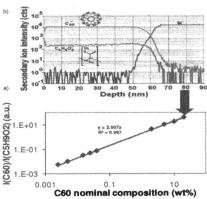

Figure 3: (a) Extinction coefficient k spectrum on PMMA:C60 films. Inset: Extinction coefficient at 261nm as a function of weight ratio of C60 introduced in solution if linear dependence is considered.

Figure 4: a) Intensity ratios obtained with ToF-SIMS plotted against expected C60 composition of the material. b) Detailed ToF SIMS ambient temperature (25°C) profile of a PMMA:C60 with 15wt% of C60.

155

In addition, thermal analyses were performed to study the physical properties of the nanocomposites. Differential Scanning Calorimetry (DSC) was used to monitor the temperatures and heat flows associated with transitions in the materials. The glass transition temperature (Tg) obtained for different C60 concentrations are reported on figure 5-a). We can point out that introduction of C60 nanomolecules in the PMMA matrix increase Tg by decreasing the intrinsic free volume. In the same direction, the thermal degradation observed by Thermo Gravimetric Analysis (TGA) is improved when nanomolecules are introduced: temperature (Td) for samples containing up to 20wt% of C60 is Td= 388°C with respect to the sample containing 0% (Td=370°C).

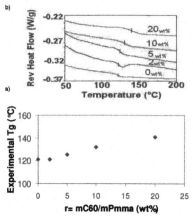

Figure 5: a)Glass Transition Temperature (Tg) obtained by Differential Scanning Calorimetry (DSC) for different material compositions. b) DSC signal for PMMA:C60 with r=0;2;5;10;20 wt%.

Figure 6: SEM image of e-Beam patterned PMMA with 10wt% of C60. 200nm pich, dose 660μC/cm2.

As the employed polymer, PMMA, is also a high resolution positive e-beam resist, we investigated the patterning capabilities of PMMA:C60 blends. In this context E-beam exposure tests were carried out on un-doped and doped PMMA samples. On figure 6 is reported the SEM picture of e-Beam patterned PMMA with 10wt% of C60 where we can clearly see the successful development of lines with ~30 nm estimated width Moreover no significant increase of the Line Edge Roughness was observed making this observation consistent with the hypothesis of single C60 or very small clusters uniformly dispersed in the matrix..

Electrical measurements

The formulated nanocomposites containing 0 to 20 Wt% of C60 were then integrated in MIS and MIM structures described on figure 1b *ii-iii* with 50 nm polymeric film thickness. The capacitance-voltage (C-V) characteristics of three different MIS devices with r= 0, 5 and 10wt% of C60 are reported on figure 7. Double sweeps were performed from inversion region to accumulation in the range of (9V,-14V, and 9V). Thermally grown SiO_2 with 5.5nm thickness was placed at the polymer/Si interface in order to suppress charge injection on the Si side.

We observe in all samples an inversion layer capacitance during the first sweep (inversion to accumulation).this effect was found to be electrode surface area and frequency dependant (data not shown here) and specific to PMMA deposition on SiO_2. We believe that this phenomenon is an edge effect related to positive charge initially present in the film (consistent with the negative Vfb observed). It could be explained by a phenomenon called "permanent inversion layer" described by Nicollian [12]: Charge of opposite sign (here positive) initially stored in the dielectric (here PMMA) can attract minority carriers at the semiconductor interface and thus provide sufficient minority carriers from capacitor edge to achieve inversion. During the reverse sweep (accumulation to inversion) the inversion level is strongly reduced. The origin of this effect is currently being investigated.

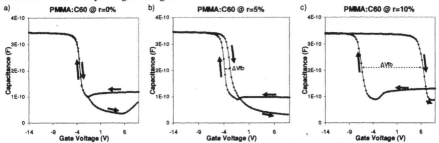

Figure 7: C-V Measurements swept from Inversion (+9V) to Accumulation (-14V) at 1 MHz for MIS capacitors with different weight ratio of C60 in PMMA (r= mC60/mPMMA). (a) r=0% - PMMA only (b) r=5% (c) r=10%.

The measurements performed on the reference device made of pure PMMA(r=0wt %) did not show any flat band Voltage (Vfb) variation between forward and backward sweeps. On the other hand, the inclusion of C60 in samples (b,c) clearly induced a clockwise hysteresis behavior that increases with C60 content. The extracted flat band voltage shift for r=5wt% and r=10wt% are respectively $\Delta Vfb_{5wt\%} = + 1.12V$ and $\Delta Vfb_{10wt\%} = + 12.85$ V, being consistent with the trapping of negative charges with C60 inclusion, as already observed in [8, 9]. Extraction of the Si/SiO_2 interface equivalent trapped charge in the case of a 10wt % of C60 was about $N = 2.8 \times 10^{12}$ cm^{-2} and it was observed that this hysteresis window could be increased by performing a sweep with larger negative voltages. Vfb was found to remain shifted after voltage sweep, confirming the interest to study retention characteristics in future work.

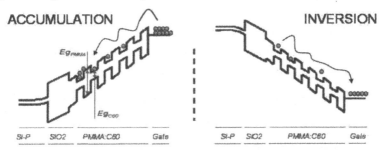

Figure 8: Schematic band structure for MIS device Si-P/Sio2/PMMA:C60/Pt in accumulation (left) and inversion (right) regimes (arbitrary thickness and band level).

The charge responsible for this phenomenon is considered as coming exclusively from gate since current blocking oxide was placed purposely on substrate side to hinder injection on that side. The mechanism could be illustrated with a band diagram on Fig 8. This band diagram was drawn approximately as HOMO and LUMO levels are not reported for PMMA (reported energy gap is EgPMMA = 5.6 eV [13]; regarding C60 energy gap is neighboring environment dependent, single C60 molecule DFT calculation report EgC60= 2.88eV [14]). In accumulation regime the electrons could flow from the gate and stay trapped within the C60. This phenomenon could be further reversed in inversion regime where electrons could be released towards the gate.

CONCLUSIONS

In summary, we evaluated here the impact of C60 molecules doping in PMMA insulating polymer. Since fullerene concentration in the material is critical to dictate the electrical behavior we demonstrated the possibility of quantifying the C60 ratio using several physical and chemical measurements, such as UV spectroscopy and ToF-SIMS. We also introduced Thermal Analysis and ToF-SIMS techniques to bring information on nanocomposite organization and distribution

For electrical characterization we suggest a MIS structure with 5.5 nm oxide at the semiconductor interface and non oxidizible electrodes to ensure a straightforward study on trapping capabilities of the considered organic nanocomposite. Based on this reference, by increasing the amount of fullerenes in the PMMA matrix, we demonstrated that C60 doping of a Poly-Methyl-MethAcrylate polymer induces electron trapping capabilities.

We also demonstrated through ebeam exposure that PMMA containing C60 doping up to 20 wt% can be patterned at low dimensions, thus highlighting that PMMA:C60 active layers for memory devices, are scalable in size.

REFERENCES

1. Y. Yang et Fred Wudl, Advanced Materials vol. 21(2009) 1401-1403.
2. J. Godlewski et M.Obarowska , Europ.Physical Journal.Special Topics 144 (2007) 51-66.
3. TW. Kim, H. Choi, SH. Oh, G. Wang, DY. Kim, H. Hwang and T. Lee, Adv. Mater. vol 21 (2009) 2497.
4. Dong Ick Son, Chan Ho You, Won Tae Kim, Jae Hun Jung, and Tae Whan Kim, Appl. Phys. Lett. 94 (2009) 132103.
5. Q-D. Ling, D-J Liaw, C. Zhuc, D. S-H Chan, E-T Kang, and K-G Neoh, Progress in Polymer Science vol. 33 (2008) 917-978.
6. S. Paul, IEEE Transactions on Nanotechnology vol. 6 (2007) 191-195.
7. H. J. Kim, J. H. Jung, J. H. Ham, and T-W Kim., Japanesee Journal of Applied Physics vol. 47 (2008) 5083-5085.
8. D-I Lee, J-H Ham, J H Jung and T-Whan Kim, Journal of the Korean Physical Society, Vol. 55, (2009), 42-45.
9. M.F. Mabrook, A.S. Jombert, S.E Machin, C Pearson,D. Kolb, Materials Science and Eng B. 159-160 (2009) 14-17.
10. M. Py, JP. Barnes, M. Charbonneau, R. Tiron, and J. Buckley Surface and Interface Analysis (2010).
11. R. Tiron et al., submitted to Microelectronic Engineering.
12. Nicollian Edward H., Brews John R., « MOS (Metal Oxide Semiconductor) Physics and Technology », New-York : J. Wiley Interscience Publication (1982).
13. J.A. Hagen, W. Li, A. J. Steckl, and J. G. Grote Appl. Phys. Lett. 88, 171109 (2006).
14. QD Ling, SL Lim, Y. Song, CX. Zhu, D.SH Chan, ET. Kang, KG. Neoh, Langmuir 23 (2007) 312-319.

Mater. Res. Soc. Symp. Proc. Vol. 1250 © 2010 Materials Research Society   1250-G04-11

Small Organic Molecules for Electrically Re-writable Non-volatile Polymer Memory Devices.

Iulia Salaoru and Shashi Paul
Emerging Technologies Research Centre, De Montfort University, Hawthorn Building,
The Gateway, Leicester, LE1 9BH,
United Kingdom
isalaoru@dmu.ac.uk
pshashi@dmu.ac.uk

ABSTRACT

The usage of organic materials in the manufacture of electronic polymer memory devices is on the rise. Polymer memory devices are fabricated by depositing a blend (an admixture of organic polymer, small molecules and nanoparticles) between two metal electrodes. The primary aim is to produce devices that exhibit two distinct electrical conductance states when a voltage is applied. These two states can be viewed as the realisation of non-volatile memory. This is an interesting development; however, there are a number of theories that have been proposed to explain the observed electrical behaviour. We have proposed a model that is based on electric dipole formation in the polymer matrix. Here, we investigate further the proposed model by deliberately creating electric dipoles in a polymer matrix using electron donors (8-Hydroxyquinoline, Tetrathiafulvalene and Bis(ethylenedithio)tetrathiafulvalene) and electron acceptors (7,7,8,8-Tetracyanoquinodimethane, Tetracyanoethylene and Fullerene) small molecules.

Two types of structures were investigated (i) a *metal/blend of polymer and small molecules/metal* (MOM), device and (ii) a *metal/insulator/blend of small molecules and polymer/semiconductor* (MIS) architecture. A blend of polymer and small organic molecules was prepared in methanol and spin-coated onto a glass substrate marked with thin aluminium (Al) tracks; a top Al contact was then evaporated onto the blend after drying - this resulted in a metal-organic-metal structure. The MIS structures consisted of an ohmic bottom Al contact, p-type Si, a polymer blend (two small organic molecules and insulating polymer), followed by polyvinyl acetate and finally a top, circular Al electrode. In-depth FTIR studies were carried out to understand the observed electrical behaviour. An electrical analysis of these structures was performed using an HP4140B picoammeter and an HP 4192A impedance analyser at a frequency of 1 MHz.

Keywords: memory devices, small organic molecules, switching mechanism

INTRODUCTION

In the recent years, these has been a growing interest in using organic materials (polymers, small molecules) in the fabrication of electronic devices such as field effect transistors, light emitting diodes and solar cells [1,2,3]. This interest is largely motivated by same important advantages of organic materials such as simplicity of the device structure; low-temperature processing; easy fabrication methods (spin-coating; drop-casting; dip-coating); and compatibility with different types of substrates (glass, plastic, metal).

Organic (polymer) memory devices (PDMs) are one of the most successful applications of organic materials. PDMs can be constituted by a large variety of materials and structures and the simplest structure is based on a single active organic layer (an admixture of a polymer and different types of the storage materials) sandwiched between two metal electrodes. Intensive research is currently focused on the storage medium such as: metal gold(Au) [4-6] or semiconducting (ZnO) [7] nanoparticles; phase change materials (Ge-Sb-Te) [8,9]; ferroelectrics (BaTiO$_3$) [10,11]; ionic complexes (NaCl) [12,13]; single wall carbon nanotubes (CNTs) [14,15]; and small organic molecules (fullerene C60; tetracyanoehylene (TCNE); tetrathiafulvalene (TTF)). Various types of small organic molecules or a combination of them with nanoparticles (Au) show great promise as charge elements in PDMs [16-23]. The first fully organic bistable memory device based on a blend of a polymer and small organic molecules (fullerene (C$_{60}$)) was reported by Kanwal et al [24]. Since then intensive research on this area has been reported by Kanwal et al. [25], Paul [26] and Yang et al [27].

In this work we investigate memory devices based on a blend of two small organic molecules where one is an electron donor (TTF, 8HQ) and another electron acceptor (TCNE, TCNQ, C60) in a polymer matrix. Dr Paul in 2007 [26] proposed a model to explain the working mechanism in PDMs based on dipoles formation in the polymer matrix and in this report we attempt to corroborate this model utilizing donor acceptor complexes.

EXPERIMENTAL

Two types of the structure were investigated in this work metal/polymer blend/metal (MOM) and metal/insulator/semiconductor (MIS) figure 1. In order to make MOM structure the organic active layer is a blend of polymer and two small organic molecules. As polymer was used polyvinyl acetate (PVAc) and as organic molecules Tetrathiafulvalene (TTF), 8-Hydroxyquinoline (8HQ), 7,7,8,8-Tetracyanoquinodimethane (TCNQ), Tetracyanoethylene (TCNE) and Fullerene (C60). The blend of polymer and small organic molecules was spin-coated onto a glass substrate marked with thin aluminium (Al) tracks; a top Al contact was then evaporated onto the blend after drying - this resulted in a metal-polymer blend-metal structure. The second structure investigated - MIS consisted of an ohmic bottom Al contact, p-type Si, a polymer blend (two small organic molecules and polystyrene), followed by polyvinyl acetate and finally a top, circular Al electrode.

160

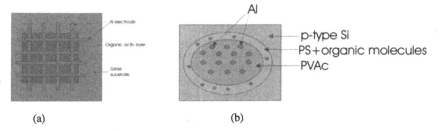

(a)   (b)

Figure 1. (a) MOM structure; (b) MIS structure

An electrical analysis of MPM and MIS structures was performed using an HP4140B picoammeter and an HP 4192A impedance analyser at a frequency of 1 MHz.

RESULTS AND DISCUSSION

Thus, aim of this work is to investigate donor-acceptor complexes embedded in the polymer matrix in the realization of PMDs. In this order the electrical behavior of metal/polymer and donor-acceptor blend /metal and metal/insulator/semiconductor structures was investigated. The structures studied was made by using a blend of two small organic molecules where one is an electron donor (TTF; 8HQ) and the second is an electron acceptor (TCNE; TCNQ; C60) and polymer as matrix. The chemical structures of all these molecules are presented in figure 2.

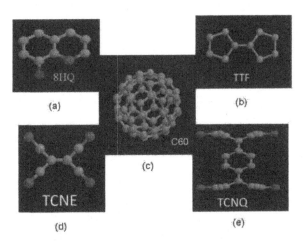

Figure 2. Chemical structures of (a) 8-Hydroxyquinoline (8HQ) (b) Tetrathiafulvalene (TTF), (c) Fullerene (C60). (d) Tetracyanoethylene (TCNE); (e) 7,7,8,8-Tetracyanoquinodimethane (TCNQ)

161

We believe that electrical behavior current-voltage (I-V) and capacitance- voltage (C-V) respectively of these devices are completely determined by the small organic molecules. In previous paper we have proposed a model to explain electrical characteristics based on electric dipole formation in the polymer matrix [10-12, 19,20,26]. In order to confirm this we used a donor acceptor molecules pairs in this work.

When these two small molecules (where one is an acceptor and another is a donor) are incorporate in the polymer matrix charge transfer occurring between them. After that one became negative charged and second one positive charged and in this way is create a dipole structure in the polymer matrix and the dipole formation lead an internal electric field. The working mechanism – switching between high and low conductivity state is completely determined by the internal electric field.

The current-voltage characteristics for five types of the donor-acceptor (D-A) complexes are shown in figure 3. It is very clear that for all complexes that we studied show two electrical conductivity states – low and high, and that switching between them is determined by the direction of internal electric field in comparative with external one. For Al/PVAc/Al device (is not shown here) the current-voltage characteristics shows a small hysteresis when voltage is applied.

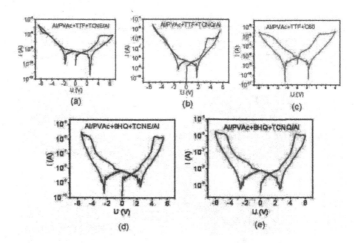

Figure 3
Current-voltage characteristics of (a) Al/PVAc+TTF+TCNE/Al; (b) Al/PVAc+TTF+TCNQ/Al; (c) Al/PVAc+TTF+C60/Al; (d) Al/PVAc+8HQ+TCNE/Al; (e) Al/PVAc+8HQ+C60/Al

Memory retention time (data not shown) was also investigated for admixture between PVAc and two small molecules (D-A complexes), with the devices exhibiting distinguishable two conductivity states (low conductivity state and high conductivity state respectively) for continuously working over a period of 4 hours.

In order to confirm that D-A complexes have a crucial role in realization of memory effect capacitance-voltage (C-V) measurements of metal/insulator/semiconductor (MIS) structure was carry out. MIS devices without small organic molecules Al/PVAc/PS/Si (p-type) show a flat hysteresis (0.2V) in the C-V characteristics (is not show here). The MIS devices which contain two small organic molecules (donor-acceptor system) show a large hysteresis. The capacitance-voltage behavior of five combination of donor-acceptor molecules (TTF+TCNE; TTF+TCNQ; TTF+C60; 8HQ+TCNE; 8HQ+TCNQ) are show in Figure 4). Thus, all these results are strong evidence of that small organic molecules plays a fundamental role in the switching mechanism in the polymer memory devices.

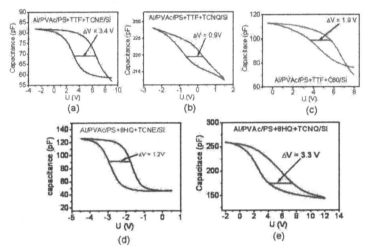

Figure 4 Capacitance-voltage characteristics of (a) Al/PVAc+TTF+TCNE/Al; (b) Al/PVAc+TTF+TCNQ/Al; (c) Al/PVAc+TTF+C60/Al; (d) Al/PVAc+8HQ+TCNE/Al; (e) Al/PVAc+8HQ+C60/Al

Additionally, it is very significant to indicate that the quality of these devices (MOM and MIS) depends strongly about the concentrations of small organic molecules and also about the molecules selection in the donor-acceptor system [20]. Also, an in-depth optical – Fourier Transform Infra-Red (FTIR) studies were carried out to understand the observed electrical behavior and will be published somewhere else.

CONCLUSION

We have successful demonstrated by current-voltage and capacitance-voltage measurements that small organic molecules have a crucial role in realization of memory effect in our metal/polymer blend/metal and metal/insulator/semiconductor structures. Thus, the working mechanism, switching between low and high conductivity state of our devices, presented here, is completely determined by the value by the internal electric field.

ACKNOWLEDGEMENT

The authors would like to thank EPSRC (Grant #EP/E047785/1) for supporting this work.

REFERENCES

[1] G. Yu, J. Gao, J.C. Hummelen, F.Wudl, A.J.Heeger, Science 270, (1995), 1789,
[2] C.W.Tang, S.A. Vanslyke, Appl. Phys. Lett. 51, (1987), 913,
[3] F. Garnier, R. Hajlaoui, A.Yassar, P. Srivastava, Science, 265, (1994), 1684.
[4] S. Paul, C. Pearson, A. Molloy, M. A. Cousins, M. Green, S. Kolliopoulou, P. Dimitrakis, P. Normad, D. Tsoukalas and M.C. Petty, Nano Lett. 3, 533-536 (2003),
[5] Prakash, A. et al., Journal of Applied Physics, 2006. 100: p.054309
[6] Möller, S., Perlov, C., Jackson, W., Taussig, C., and Forrest, S.R, Nature 426, 166-169, (2003)
[7] A.N.Aleshin, E.L.Alexandrova, Physics of the Solid State 50(10) (2008) 1895.
[8] M.H.R. Lankhorst, B.S.S.S.Ketelaars and R.A.M.Wolters, Nature materials, 4, 347, (2005)
[9]W.Welnic, M.Wuttig, Materials Today, 11(6), (2008)
[10] I.Salaoru, S.Paul, Journal of Optolelectronics and Advanced Materials 10(12), 3461, (2008).
[11] I.Salaoru, S.Paul, Phil.Trans.R.Soc.A, 367, 4227, (2009).
[12] I.Salaoru, S.Paul, Advances in Science and Technology 54, 486 (2008).
[13] J.H.Krieger, S.V.Trubin, S.B.Vaschenko, N.F.Yudanov, Synth.Met.,122, (2001), 199
[14]X.B.Lu, J.Y.Dai, Appl.Phyl.Lett., 88, (2006), 113104
[15]M.F.Mabrook, A.S.Jombert, S.E.Machin, C.Pearson, D.Kolb, K.S.Coleman, D.A.Zeze, M.C.Petty, Materials Science and Engineering B, 159-160 (14), (2009),
[16] H.J.Kim, J.H.Jung, J.H.Ham and T.W.Kim, Japanesse Journal of Applied Physics, 47(6) (2008) 5083.
[17] J.H.Ham, J.H.Jung, H.J.Kim, D.U.Lee and T.W.Kim, Japanesse Journal of Applied Physics 47(6) (2008) 4988.
[18] C.W.Chu, J.Ouyang, J.H.Tseng, Y.Yang, Adv.Mater. 17 (2005) 1440.
[19] I.Salaoru, S.Paul, Mater.Res.Soc.Symp.Proc, 1114-G12-09, (2009),
[20] I.Salaoru, S.Paul accepted to Thin Solid Films
[21] J.Y. Ouyang et. al., Appl.Phys.Lett. 86 (2005) 123507.
[22] D.Prime, S.Paul, Appl.Phys.Lett., 96, (2010), 043120,
[23] D.Prime, S.Paul, Mater. Res. Soc. Symp. Proc. 0997-I03-01 (2007).
[24] A.Kanwal, S.Paul and M.Chhowalla, MRS Proc. 830 (2005) 349.
[25] S.Paul, A.Kanwal, M.Chhowalla, Nanotechnology 17 (2006) 145.
[26] S Paul, IEEE Transactions on Nanotechnology 6 (2007) 191.
[27] Y.Yang, J.Ouyang, L.Ma, J.H.Tseng, C.W.Chu, Adv. Funct. Mater. 16 (2006) 1001.

Phase Change RAM

Mater. Res. Soc. Symp. Proc. Vol. 1250 © 2010 Materials Research Society   1250-G14-01-H07-01

Phase Change Memory with Chalcogenide Selector (PCMS): Characteristic Behaviors, Physical Models and Key Material Properties

I. Karpov, D. Kencke, D. Kau, S. Tang*, and G. Spadini
Intel Corporation, 2200 Mission College Blvd., Santa Clara, California 95054
*Numonyx B.V. , 2550 N. 1st Street, Suite 250, San Jose, CA 95131

ABSTRACT

We present a novel scalable and stackable nonvolatile solid state memory. Each cell consists of a storage element, based on phase change memory (PCM) element, and an integrated selector, using an Ovonic threshold switch (OTS). The cell is implemented to enable a true cross-point array. The main device characteristics and behaviors, corresponding physical processes in different operation modes, and key material properties are discussed.

INTRODUCTION

The chalcogenide phase change device utilizes electrically initiated reversible amorphous-to-crystalline phase changes in multi-component chalcogenides, such as $Ge_2Sb_2Te_5$, whose significantly different phase resistances are used as the two logic states [1]. For each PCM element in the array a selector element is required to avoid a parasitic path in the resistive network In this work, a thin-film two-terminal threshold switch, OTS, [2,3] is used as the selector for PCM arrays [4]. In contrast to PCM, the composition of chalcogenide material in the switch is selected [2,3] to maintain its amorphous/glassy state regardless of applied pulse characteristics. The switch is integrated with a PCM element to form the PCMS device and isolate individual PCM cells in the cross point array.

EXPERIMENT

PCMS cell and memory array are shown in Fig. 1a and 1b, respectively. The vertical stack of a PCMS cell consists of OTS and PCM elements interlinked by a middle electrode, with a top electrode connecting OTS to bit line (column) above and a bottom electrode contacting PCM below to the word line (row). The memory cell stack, including rows and columns, can be sandwiched between back end layers and fully integrated with a CMOS technology.

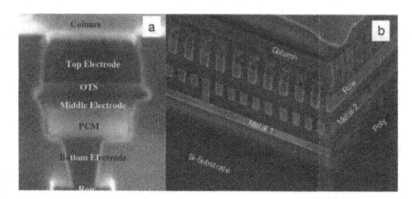

Fig. 1. (a) SEM of PCMS cell. (b) PCMS cross-point array, sandwiched between M2 and M3 metal layers in the back end of the process, and integrated with CMOS technology [4]

As illustrated in Fig. 1(a) PCMS cell consist of two devices, PCM and OTS, in series. These devices may be fabricated for characterization by leaving out either the PCM or the OTS layer. The *I-V* characteristics of individual PCM and OTS cells, fabricated by dedicated process flows, are shown in Fig. 2(a) and 2(b), respectively. Note that a SET or RESET programming pulse changes the memory state of the PCM device but does not change IV characteristic of the OTS device.

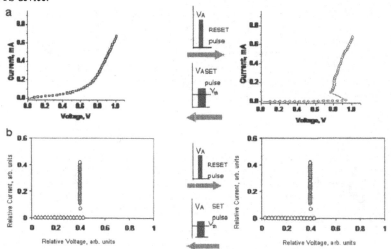

Fig. 2. (a) Typical *I-V* characteristics of a dedicated PCM cell in SET (left) and RESET (right) states. (b) Typical *I-V* characteristics of the OTS cell after experiencing SET (left) and RESET (right) pulses.

PCMS cell *I-V* characteristics are shown in Fig. 3. SET and RESET states have different threshold voltages. The SET state has lower threshold voltage than the RESET state. Two different threshold voltages, high and low, correspond to two logic states, 1 and 0.

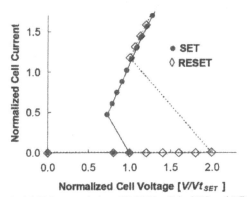

Fig. 3. Typical *I-V* characteristics of PCMS cell in SET and RESET states.

The programming transfer characteristics for PCMS cell are best described by a V_t-I curve because each state of the cell is characterized by threshold voltage, V_t, instead of by resistance, R, as has been previously described in R-I curves [1]. A typical V_t-I characteristic for PCMS is illustrated in Fig. 4. When programming pulse amplitudes are low, the PCM element is still in a crystalline state and the V_t of PCMS cell is low and is determined by the threshold of the OTS device. The higher V_t at higher programming current is due to formation of an amorphous phase in PCM and due to the serial connection of two amorphous alloys, OTS and amorphous PCM.

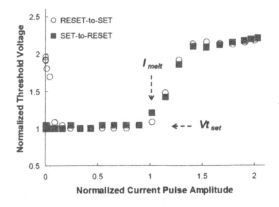

Fig. 4. Typical programming characteristics of PCMS cell.

169

DISCUSSION

Fig. 5 shows distribution of threshold voltages in PCMS devices in a 2-Mb array block demonstrating a 1-V V_t window. The mechanism of threshold switching, the different features it

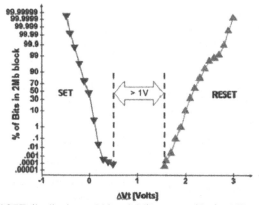

Fig. 5. RESET and SET distributions within a 2-Mb memory block. ΔVt on the x-axis refers to the mean V_t of SET state.

exhibits, and its dependence on ambient temperature, device geometry and material parameters are all key to PCMS technology development. Fig. 6(a) shows the temperature dependence of threshold voltage V_t, which is lower for higher temperatures. Fig. 6(b) shows thickness dependences of V_t for both programmed and all amorphous GST thin film devices based on the lance cell architecture [5]: the measured V_t increases with thickness. The linear dependence of V_t on thickness points at threshold initiated by a certain critical field.

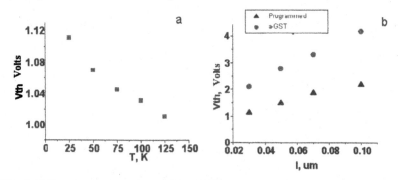

Fig. 6. (a) Temperature dependence of threshold voltage for GST. (b) Threshold voltage dependence on GST layer thickness

The design of the PCMS cell affects how V_t depends on geometrical dimensions. Fig. 7 represents PCM simulation [6] that showing how a misalignment of bottom electrode contact changes threshold voltage for different levels of programming current. Offset of the bottom electrode contact away from center redirects more current toward the center of the device. It produces more heat there that somewhat compensates for the offset. Once enough current is supplied, the amorphized region is similar in size and V_t is less sensitive to the offset.

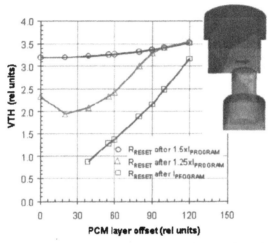

Fig. 7. Simulated threshold voltage dependence on bottom electrode contact misalignment for different levels of programming current in a PCM cell configured as in Fig. 1(a). Simulations approximated the cell as a 2D cross-section. Results were also compared a full 3D simulation (inset).

In addition to V_t vs. temperature and cell geometry, it is important to look at its temporal dependence. We observed that after the voltage across the amorphous phase grows above V_t, the device can idle for a certain 'delay' time τ_d and then switch to a low resistive ON state as shown in Fig. 8(a). We measured variation in switching delay time, τ_d, and in V_t for the same cells and found that variation in delay time could be significant with a standard deviation reaching 40%. But this variation was found to decrease rapidly when the amplitude of the applied voltage is increased as shown in Fig. 8(b) [6].

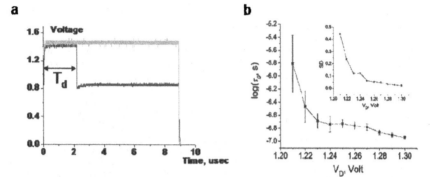

Fig. 8. (a) Example of the oscilloscope traces of applied and device voltages showing how switching delay time is measured. (b) Switching delay time and its variation vs. applied voltage for GST cell

A unified field-induced nucleation model [8] provides a common mechanism and analytical description for switching in both PCM and OTS devices of arbitrary thickness. The model is solved analytically to quantify the relationship between V_t and material parameters, such as the nucleation barrier and radius, amorphous layer thickness, as well as its dependence on temperature [9]:

$$V_t = \frac{Wo}{kT} \frac{V_{\min}}{\ln\left(\frac{t}{\tau}\right)}, \quad V_{\min} = l\sqrt{\frac{6 \cdot Wo \cdot \alpha^3}{\varepsilon Ro^3}} \quad (1)$$

where Wo and Ro are the nucleation barrier and radius, ε is the dielectric permittivity, τ is the characteristic time of phase transformation, t is the switching delay time, related to the incubation time for a needle-like embryo triggering the formation of a crystalline shunt, $\alpha \sim 0.1$ is a numerical parameter. We note that value of $2 \cdot Ro$, ~5nm, determines the minimal dimension for the active volume in working PCM cell. The nucleation-switching model was also extended recently to describe statistics of threshold voltage and switching delay time [10]. As applied to both PCM and OTS, it introduces the concept of holding voltage Vh and relation between Vth and Vh, both expressed through material parameters [8,11], of which the nucleation barrier and radius appear to be most important.

Another observed temporal variation in threshold voltage depends on the 'wait' time, t_{wait}, between the time the device was programmed and the time of measurement of V_t [12]. Dependence V_t and R, measured for fully RESET device, on t_{wait} is shown in Fig. 9.

172

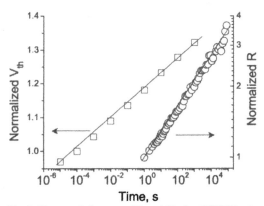

Fig. 9. Temporal changes of R and V_t after RESET pulse for GST

This behavior was analyzed in the framework of atomic transformation in disordered systems [12]. Due to dramatically fast quench rate, $\sim10^{10}$ K/s, the atomic subsystem does not have enough time to settle into thermodynamically stable state. Therefore the metastable RESET state relaxes to its more stable state with time. This phenomenon was analytically described in terms of random atomic double well potentials (DWP) representing a metastable disordered atomic structure in a glass with fluctuations of atomic barrier heights to show that

$$R(t) = R_0(t/t_0)^\alpha \quad (2) \qquad \frac{\Delta V_t(t)}{V_t(0)} = v\ln(\frac{t}{t_o}) \quad (3)$$

where v is threshold voltage drift coefficient and α is resistance drift coefficient related to material properties such as deformation potential.

It is a difficult task to list all physical phenomena underlying the operation of PCMS devices. Nevertheless Fig. 10 shows a diagram illustrating how three main device operations, write SET, write RESET, and read are related to different physical processes in the cell.

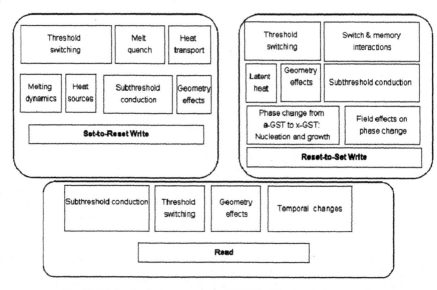

Fig. 10. Main physical processes in PCMS cell during device operation

Material properties of different components of PCMS cell are important for reliable device operation. In Fig. 11 we highlight main material properties of chalcogenide materials which are needed to successfully model operation of PCMS device.

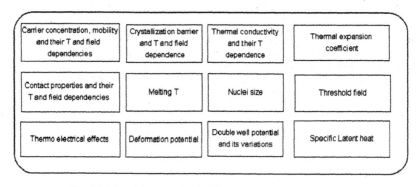

Fig. 11. Material properties for simulation of PCMS device operations

CONCLUSIONS

An Ovonic threshold switch has been integrated with a PCM element to create a PCMS cell, allowing area-efficient memory layout for future solid-state storage applications. Optimization of material properties in the PCMS cell is a key enabler for this promising NVM technology.

REFERENCES

1. Stefan Lai and Tyler Lowrey, *IEDM Tech. Dig.* (IEEE, 2001), p. 903.
2. S. Ovshinsky, *Phys. Rev. Lett.*, **21**, p.1450 (1968).
3. D. Adler, H. Henisch, and N. Mott, *Reviews of Modern Physics*, **50**, p. 209 (1978)
4. D. Kau, S. Tang, I. Karpov, R. Dodge, B. Klehn, J. Kalb, J. Strand, A. Diaz, N. Leung, J. Wu, S. Lee, T. Langtry, K. Chang, C. Papagianni, J. Lee, J. Hirst, S. Erra, E. Flores, N. Righos, H. Castro, and G. Spadini, *IEDM Tech. Dig.* (IEEE, 2009), p. 617.
5. I. Karpov, S. Savransky, and V. Karpov, *22nd IEEE Nonvolatile Semiconductor Memory Workshop*, (IEEE, 2007), p. 56.
6. D. Kencke, I. Karpov, B. Johnson, S. Lee, D. Kau, S. Hudgens, J. Reifenberg, S. Savransky, J. Zhang, M. Giles, and Gianpaolo Spadini, *IEDM Tech. Dig.* (IEEE, 2007), p.323.
7. I. Karpov, M. Mitra, D. Kau, G. Spadini, A. Kryukov, and V. Karpov, *Appl. Phys. Lett.* **92**, 173501 (2008).
8. M. Nardone, V. Karpov, D. C. S. Jackson, and I. V. Karpov, *Appl. Phys. Lett.* **94**, 103509 (2009).
9. V. Karpov, Y. Kryukov, I. Karpov, and M. Mitra, *Phys. Rev. B*, **78**, 052201 (2008)
10. V. Karpov, Y. Kryukov, M. Mitra, and I. Karpov, *J. Appl. Phys.* **104**, 054507 (2008).
11. M. Nardone, V.G. Karpov, and I. V. Karpov, *J. Appl. Phys.*, **107**, 054519 (2010).
12. I. Karpov, M. Mitra, D. Kau, G. Spadini, Y. Kryukov, and V. Karpov, *J. Appl. Phys.* **102**, 124503 (2007) and I. Karpov, M. Mitra, D. Kau; G. Spadini, Y. Kryukov, and V Karpov, *International Symposium on VLSI Technology, Systems and Applications* (IEEE, 2008). p.140.

Mater. Res. Soc. Symp. Proc. Vol. 1250 © 2010 Materials Research Society   1250-G14-04-H07-04

Memory retention of doped SbTe phase change line cells measured isothermally and isochronally

J.L.M. Oosthoek[1], B.J. Kooi[1], K. Attenborough[2], G.A.M. Hurkx[2], D.J. Gravesteijn[2]
[1]Materials Innovation Institute M2i and Zernike Institute for Advanced Materials, University of Groningen, Nijenborgh 4, 9747 AG, Groningen, The Netherlands.
[2]NXP-TSMC Research Center, Kapeldreef 75, 3001 Leuven, Belgium.

ABSTRACT

Doped SbTe phase change (PRAM) line cells produced by e-beam lithography were cycled for at least 100 million times. The memory retention of the PRAM cell was measured both isothermally and isochronally which showed excellent agreement. An activation energy for growth of 1.7 eV was found (after 100 million cycles) for both measurements. Similar isothermal and isochronal measurements were performed on PRAM cells produced by optical lithography which yielded activation energies of 3.0 eV and 3.3 eV, respectively. Our results show that the same phase-change material can show large differences in retention behavior depending on the way the cells are produced.

INTRODUCTION

Currently, Flash memory is the technology of choice for non-volatile memory applications. Although it already exceeds the expectations that were foreseen in the past, it is expected that down-scaling will become increasingly difficult. Phase-change random access memory (PRAM) is a potential candidate to replace Flash memory in the near future. PRAM expected to be scalable with the next generations of lithography, it requires less lithographic steps, it has a much higher writing speed and is more energy efficient to program.[1-3]

The information carrier in a PRAM cell is a nano-sized resistor made from phase-change material. Information is stored as a difference in electrical resistance between the amorphous and crystalline phase which is typically three orders of magnitude. It can be switched electrically between the two phases reversibly. By supplying a short high energy (RESET) pulse the crystalline resistor is melted and, when cooled sufficiently fast, quenched into the amorphous phase. By supplying a longer but lower energy (SET) pulse the temperature is raised above the glass temperature but below the melting temperature. This allows crystal growth into the amorphous mark, which can fully crystallize in the order of a hundred nanoseconds.[1-2] The amorphous phase is meta-stable by nature. Over time the cell returns to the crystalline phase (SET state) by spontaneous crystal growth and nucleation of the phase-change material which is by all means an undesirable effect. These processes, like most phase transformations, are thermally activated, because the underlying physical phenomena such as atomic jump frequencies, viscosity, diffusion, nucleation and growth are thermally activated.[4] This thermal activation is generally considered to be of Arrhenius type.

Currently, in many applications so-called fast-growth type phase-change materials are employed, where the crystallization process only depends on growth and not on nucleation. One such an application is in PRAM line cells.[2] As the crystal grows from the already present crystal boundary the growth process is in principle one-dimensional. Because the crystal growth rate in PRAM line cells is not easily obtained, the retention time, i.e. stability of the amorphous mark at a fixed temperature, is measured instead. The retention time is generally inversely proportional to the growth rate.

Instead of this isothermal analysis, more convenient and popular is to perform isochronal measurements, i.e. measuring the stability of the amorphous mark with various heating rates, kept constant in time, where the temperature at which the cell crystallizes is the parameter being measured. *Only* in the 1D-growth case of the line cells, an exact relation can be derived between the retention times of the cells measured isothermally and isochronally.

The activation energy can then be determined from the data of which the so-called Kissinger analysis[5], which is a (first-order) approximate method, is best known. Accurate memory retention measurements are slow, because the most significant data points are closest to operating temperatures and retention times, for example 10 years at 80°C. In contrast to isothermal measurements, isochronal measurements have the advantage that an upper limit of how long a measurement is going to take is better known a-priori.

In the present work we show results on cyclability, amorphous resistance drift and data retention of PRAM line cells and particularly demonstrate that the retention behavior as measured isochronally with the Kissinger method excellently agrees with retention behavior measured isothermally.

THEORY

We start with the assumption (which can be validated experimentally) that the growth rate G follows an Arrhenius dependence:
$$G_T = G_\infty \cdot e^{-Eg/kT} \tag{1}$$
Then, in order to determine the activation energy, e.g. for crystal growth, the most straight-forward and accurate procedure is to determine the growth rate at various temperatures, and then to plot the logarithm of growth rate versus the reciprocal (absolute) temperature (T). The slope from a linear fit to the data is then equal to the activation energy (E_g) divided by Boltzmann's constant (k). In PRAM line cells crystal growth occurs at both amorphous-crystalline boundaries.[6]

For a line cell with an amorphous mark having a length of 2L, the retention time τ_T can therefore be taken directly from Eq.1 as:
$$\tau_T = L/G_T = \tau_\infty \cdot e^{Eg/kT} \quad (\text{with } \tau_\infty = L/G_\infty) \tag{2}$$
The activation energy for growth can be directly obtained from retention times. Then, in order to derive the isochronal case, the Arrhenius equation is integrated assuming a constant increase of temperature as a function of time: $T = \varphi \cdot t$ i.e. with a constant ramp rate φ. The lower temperature, where the integration starts is not critical, because only the temperatures close to the crystallization temperature (T_c) have a relevant influence on the integral. The distance dL a crystal grows at a certain time dt will be $dL = G_T dt = (G_T/\varphi) \cdot dT$. The length L a crystal boundary has grown at a given temperature with a given ramp rate will thus be:
$$L = \int \frac{G_\infty}{\varphi} exp(\frac{Eg}{kT}) dT \tag{3}$$
The solution of this integral, running from 0 to T_c, can be represented as an infinite power series:
$$L = \frac{G_\infty}{\varphi} exp(\frac{Eg}{kT_c}) \cdot \sum_{i=1}^{\infty} i! T_c (T_c k/E_g)^i \tag{4}$$
From the first-order approximation the usual Kissinger equation can be derived:
$$L = \frac{G_\infty}{\varphi} exp(\frac{Eg}{kT_c}) \cdot T_c^2 k/E_g \implies ln(\varphi/T_c^2) = ln(LE_g/kG_\infty) - E_g/kT_c \tag{5}$$

By plotting $\ln(\varphi/T_c^2)$ versus $1/kT_c$ the slope clearly yields the value $-E_g$ which is the original Arrhenius activation energy. The intercept of $\ln(\varphi/T_c^2)$ versus $1/kT_c$ yields $\ln(LE_g/kG_\infty)$. The Arrhenius pre-factor $\tau_\infty=L/G_\infty$ is taken from the intercept using the value of E_g from the slope. The Kissinger analysis can be extended to higher order (or more) approximations. For three orders, the activation energy E_g is obtained from the slope of $\ln(\varphi/(T_c^2-2kT_c^3/E_g+6k^2T_c^4/E_g^2))$ versus $1/kT_c$. The fact that E_g is obtained from data points that already contain E_g is solved iteratively. The pre-factor is found similarly.

It is important to notice that the Kissinger plot *requires* that retention follows the Arrhenius behavior. If an Arrhenius type temperature dependence behavior does not occur then the Kissinger analysis does not have a physical relevance.

EXPERIMENT

Cyclability

Doped SbTe line cells with dimension $225\times50\times20$ nm^3 were produced by e-beam lithography; details can be found in ref. 7. A cell was cycled with a series of RESET/SET pulses (see figure 1). It was shown that 100 million cycles were possible before the cell becomes stuck in the SET state; see Fig.1. Typically, during the first few hundred cycles the cell does not show a constant behavior. This initialization phase can be related to residual effects from cell production: e.g. initial stresses and trapped argon atoms from sputter-deposition. The useable life span ranges over 10 millions cycles during which the resistance difference between the crystalline and amorphous state is at least 3 orders of magnitude. The end of life is characterized by a large drop in amorphous resistance until the cell becomes stuck in the SET state. The cell can still be brought to the RESET state but the magnitude of the switching current has to be increased. The resistance of the crystalline state also drops slightly. Although this drop is small compared to the drop in amorphous resistance, it is significant and may be related to changes in atomic (de)composition of the programmable region.

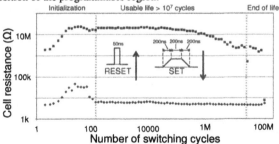

Figure 1. *A phase change line cell was cycled (RESET/SET) for 100 million times. RESET pulses of 1.1 mA, 50 ns pulse width with 4 ns edges and a SET pulses of 500 µA, 200 ns pulse width with 200 ns edges were used.*

Resistance drift

After cycling a 100 million times, the PRAM line cell was brought to the RESET state with a 50ns/1.1mA pulse at a temperature of 25.0 ± 0.1 °C. The resistance was measured for 1000 seconds with a source meter (at 100mV) starting directly after the RESET pulse. Results are shown in the inset of Fig.2a. The resistance drift as a function of time follows a well known power law $R=R_1\cdot t^\alpha$ where R_1 is the resistance at one second after RESET and α is a power law

179

exponent[8]; see the inset in Fig.2a. A value of $(50\pm5)\cdot10^{-3}$ was found for α., which is comparable to a value of 0.06 found earlier.[8] After this drift period either an isothermal or isochronal retention experiment was started. The temperature was measured with a thermocouple located very close to the cell and kept stable with a PID controller connected to a linear power supply and heating filament. After the cell was crystallized during the experiment it was brought back to 25.0°C and the whole sequence of RESET, drift was repeated and a new isothermal or isochronal retention experiment was started.

Isochronal retention

The temperature is raised with a constant ramp rate φ (in °C/minute) until the cell crystallizes; see Fig.2a. As the temperature is raised the resistance drops steadily. This initial drop in resistance, which is not dependent on the ramp rate (see Fig.2a), cannot be attributed to crystal growth but to the natural temperature dependence of the amorphous low field resistance (R_T) which follows the Arrhenius equation $R_T = R_\infty \cdot e^{Ec/kT}$ where R_∞ is a pre-factor and E_c is the activation energy for conduction. The value of E_c is generally half the value of the band gap.[8] A value of 0.29 ± 0.01 eV was found here, which is comparable to values found in the literature[8] even though a different material and cell geometry was used. The natural strong dependence of the cell resistance on temperature was conveniently used to verify that there is no (ramp rate dependent) thermal lag between the cell and the measured temperature for the ramp rates that were used.

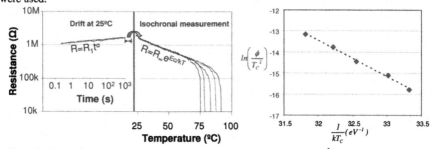

Figure 2a. For each isochronal measurement (see text), the cell is first drifted for 10^3 seconds to avoid influence of drift on the temperature measurement. The temperature is increased with a constant ramp rate and the resistance is measured until the cell is crystallized. The crystallization temperatures (T_c) from 2a are shown in a Kissinger plot in figure 2b. The resistance drift in figure 2a was reproduced for each measurement proving that the cell returned to the same state for all isochronal measurements.

Crystallization occurs when the resistance starts to deviate from this temperature dependent thermally activated resistance. The crystallization temperature (T_c) is taken at the point with the maximum drop in resistance (on a logarithmic scale). This is the moment when the crystalline boundaries (of the amorphous mark) meet and a continuous crystalline conductive path is formed.[6] The crystallization temperatures are shown in a Kissinger plot depicted in Fig.2b. From the data an (equivalent Arrhenius) activation energy for crystal growth is found of 1.7 eV and a pre-factor of $6.4\cdot10^{-23}$ s (shown in figure 3b as a red line).

Isothermal retention

The temperature was raised to a certain temperature and the resistance was measured until the cell resistance dropped below 10 kΩ. The time between the moment that the

180

temperature was reached and the time that showed (within a time step) the largest drop in resistance (on a logarithmic scale) is taken as the value of the retention time. The transition generally occurs very fast; see Fig.3a. This can be explained by a gradual growth from the crystal boundaries that suddenly meet at this point, possibly assisted by the voltage of resistance measurement. Retention times were measured ranging over almost five orders of magnitude; see Fig.3b.

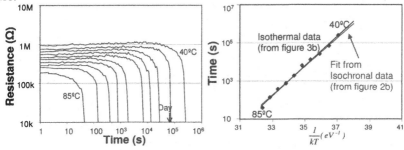

Figure 3a. After a RESET pulse and drift period (see text), the resistance is measured until the cell is crystallized. The memory retention data of figure 3a is shown in an Arrhenius plot in figure 3b. The red line in figure 3b is a fit from equivalent Arrhenius activation energy and pre-factor obtained from the isochronal measurement shown in figure 2b. It is shown that the isothermal and isochronal data are in excellent agreement.

The data clearly show that an Arrhenius type retention behavior is met over the whole range. Also, the Arrhenius retention that was calculated from the Kissinger data (red line in figure 3b) shows excellent consistency with the actual Arrhenius data, demonstrating that for these cells the two measurement techniques are interchangeable. Activation energies for growth of 1.7 eV were found for both isothermal and isochronal measurements. Although the data are very consistent, the retention behavior does not meet the requirements from industry.

The measurement was repeated on 700x300x20 nm³ PRAM cells of identical composition but produced by optical lithography; details can be found in ref 9. Activation energies for growth of 3.0 eV and 3.3 eV were found for isothermal and isochronal measurements respectively, which are consistent within measurement limits (see figure 4). These values are in excellent agreement with earlier measurements on phase-change blanket films of identical composition and layer thickness.[10] Small differences in activation energy lead to larger differences in the (extrapolated) retention at operating temperatures. The isothermal measurements indicate memory retention of 10 years at 89 °C, isochronal indicates 10 years at 92 °C.

Interesting large differences are found between the retention behavior obtained for the cells produced by e-beam or optical lithography. For the e-beam cells, we have proof that the memory retention degrades with cycling: The crystallization temperature (measured with a ramp rate of 30 °C/minute) decreases from 122 °C to 92 °C during the 10^8 cycles.[11] Also, the activation energy decreases with cycling from 2.2 ± 0.2 eV to 1.7 ± 0.2 eV. Furthermore, the smaller geometry of the e-beam cells predicts slightly lower retention. Therefore, cell cycling and geometry influences memory retention but cannot explain the large differences between cells produced by e-beam and optical lithography on its own. These results demonstrate that the same phase change material can exhibit large differences in properties such as crystallization temperature and activation energy depending on the way it is packed and processed in the device.

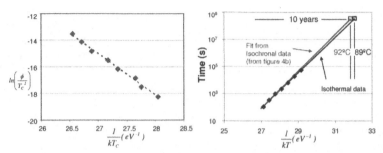

Figure 4a. The memory retention of a lithography produced cell is measured isochronically. These data are compared to an isothermal measurement of retention in figure 4b and show excellent agreement.

CONCLUSIONS

Doped SbTe PRAM line cells with dimensions 225x50x20 nm³ produced by e-beam lithography could be SET/RESET for at least 100 million cycles. The memory retention of the PRAM cell was measured both isothermally and isochronally, which gave the same activation energies for growth, 1.7 eV. The isothermal and isochronal retention measurements were repeated on PRAM cells produced by optical lithography and activation energies of 3.0 eV and 3.3 eV were found, respectively. Although the isothermal and isochronal retention measurements were again very similar, they strongly deviate from what we found for the cells produced by e-beam lithography. This deviation mainly originates from the different ways the cells are produced. Cell cycling and geometry explained for a smaller part the differences that were found.

EXPERIMENT

The research was carried out under project number MC3.05241 in the framework of the Strategic Research program of the Materials innovation institute M2i, (the former Netherlands Institute for Metals Research or NIMR). Financial support from the M2i is gratefully acknowledged.

REFERENCES

1. L. Lacaita, Solid-State Electron. **50**, 24 (2006)
2. M. H. R. Lankhorst *et al.*, Nat. Mater. **4**, 347 (2005)
3. S. Hudgens, B. Johnson, MRS Bulletin **29**, 829-832 (2004)
4. D.A. Porter, K.E. Easterling, *Phase Transformations in Metals and Alloys*, 2nd Ed. (CRC Press, 1992)
5. H.E. Kissinger, Anal. Chem. **29**, 1702 (1957)
6. J.L.M. Oosthoek *et al.*, E\PCOS (2009) Conference proceeding (www.epcos.org)
7. F. J. Jedema *et al.*, Appl. Phys. Lett. **91**, 203509 (2007)
8. A. Pirovano *et al.*, IEEE Trans. Elect. Devices **51**, NO. 5 (2004)
9. D.T. Castro *et al.*, IEEE Int. Elect. Devices Meeting, 315-318. Washington, DC, (2007)
10. J.L.M. Oosthoek *et al.*, Microscopy and Microanalysis. **16** (2010) (Accepted)
11. J.L.M. Oosthoek *et al.*, (Under preparation)

Mater. Res. Soc. Symp. Proc. Vol. 1250 © 2010 Materials Research Society   1250-G15-02

Nonvolatile Floating Gate Memory Devices Containing AgInSbTe-SiO$_2$ Nanocomposite Thin Film Prepared by Sputtering Method

Kuo-Chang Chiang and Tsung-Eong Hsieh
Department of Materials Science and Engineering, National Chiao Tung University,
1001 Ta-Hseuh Road, Hsinchu, Taiwan 30010, R.O.C.

ABSTRACT

AgInSbTe (AIST)-SiO$_2$ nanocomposite layer prepared by a one-step sputtering process utilizing target-attachment method was implanted in the nonvolatile floating gate memory (NFGM) devices. Device sample subjected to post annealing at 400°C for 2 min in atmospheric ambient exhibited a significant hysteresis memory window (ΔV_{FB}) shift = 5.91V and charge density = 5.22×10^{12} cm^{-2} after ±8V voltage sweep. During the retention time test, a ΔV_{FB} shift about 3.50 V and charge loss about 28.4% were observed in the sample after a ±5V voltage stress for 10^4 sec. Cross-sectional TEM revealed that the nanocomposite layer contains the crystalline AIST nanoparticles with the sizes about 5 to 7 nm embedded in SiO$_2$ matrix. XPS analysis indicated that annealing induces the reduction of antimony oxides to form metallic Sb nanocrystals and suppresses the oxygen defects and charge loss in nanocomposite layer. Analytical results illustrated that the utilization of AIST-SiO$_2$ nanocomposite layer may simplify the preparation of NFGM device with satisfactory electrical properties, implying a promising feasibility of such a nanocomposite layer to NFGM devices.

INTRODUCTION

In recent years, NFGM devices have received a vast attention for nonvolatile flash memory applications. In contrast to conventional device structure, NFGM device containing Si nanocrystals as the charge storage traps was first reported by Tiwari et al.[1] Such a device possesses the advantages including lower consumption, higher operative efficiency and better endurance.[2] Since then, semiconductor nanocrystals such as Si[3], SiGe[4] and Ge[5], transition metal nanocrystals such as Ag[6], Au[7], Ni[8], Pt[9] and W[10] have been implanted in NFGM devices. Among these, transition metal nanocrystals have attracted numerous research interests due to their advantages of high charge-storage traps, high charge density around Fermi level, high work functions with wide range of selection, small energy disturbance and strong coupling with the conduction channel.[6-11] However, NFGM containing metallic nanocrystals suffer from the difficulties of metal diffusion and oxidization during device preparation. Capping the metallic nanocrystals by dielectric materials to inhibit oxidation and improve the leakage current and device endurance is frequently adopted to overcome these difficulties. However, it invokes the difficulties in controlling interface compatibility and the oxide layer thickness. Wang *et al.*[12] reported that the nitrogen incorporation may drastically reduce the leakage current and escalate the breakdown fields by suppressing oxygen vacancy diffusion through the gate dielectrics. In addition, annealing in specific gas ambient at temperatures above 600°C is necessary for samples containing SiO_x matrix.[13] New materials and simple fabrication process in conjunction with low-temperature thermal treatment are hence required for the development of NFGM.

Chalcogenides are well-known phase-change media for optical recording due to their ultra fast recrystallization rate and comparatively low recrystallization temperature.[14] It is also found that work functions of Sb (4.55-4.7 eV) and Te (4.95 eV) are similar to those of transition metals.[15] Further, it has been demonstrated that AIST-SiO$_2$ nanocomposite layer can be easily prepared by using target-attachment method[16] and composite target sputtering method[17]. Owing to these findings, we prepared the NFGM device containing such a nanocomposite layer and investigated its feasibility to NFGM devices.

EXPERIMENTAL DETAILS

AIST-SiO$_2$ nanocomposite layers about 35 nm in thickness were deposited on the *p*-type, Si(100) substrates by target-attachment method[16] in a sputtering system at background pressure $\leq$ 1×10^{-6} torr. The deposition was achieved at RF power = 100W and inlet Ar/N$_2$ gas flow ratio = 100:1. Subsequently, a post annealing treatment was performed at 400°C for 2 min in atmospheric ambient to induce the recrystallization of AIST nanoparticles. Afterward, Al electrodes were deposited on nanocomposite layer by *e*-beam evaporation to complete the metal-oxide-semiconductor device preparation. Capacitance-voltage (C-V) and retention time properties were measured at 1 MHz to evaluate the nonvolatile memory characteristics of devices. Microstructure and chemical status of elements in AIST-SiO$_2$ nanocomposites were analyzed by transmission electron microscopy (TEM, JEOL-2100F) equipped with energy dispersive spectroscopy (EDX, Link ISIS 300) and *x*-ray photoelectron spectroscopy (XPS, PHI Quantera SXM), respectively.

RESULTS AND DISCUSSION

Nonvolatile Memory Properties

Figure 1(a) presents the C-V profiles of 400°C-annealed sample subjected to ±3V to ±8V gate voltage sweep. A significant memory window (ΔV_{FB}) shift can be seen and, according to the formula proposed by Maikap *et al.*[18], the maximum storage charge density is 5.22×10^{12} cm^{-2} according to the ΔV_{FB} shift = 5.91V at ±8V voltage sweep as shown in Fig. 1(a). Further, the ΔV_{FB} shift toward negative bias side is ascribed to the presence of oxygen vacancies and/or positive fixed charges in nanocomposite layer. Hysteresis loops for as-deposited and

$400^{\circ}C$-annealed samples are given in Fig. 1(b). As-deposited sample exhibits a ΔV_{FB} shift = 2.65V and the charge density = 2.39×10^{12} cm^{-2} after $\pm 8V$ voltage sweep, which is inferior to those of annealed samples. This illustrates the annealing treatment benefits the charge-storage effect in nanocomposite layer and thus improves the nonvolatile memory characteristics of NFGM devices.

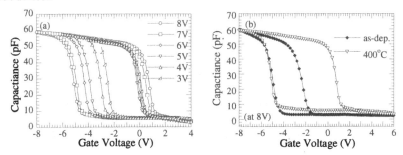

Figure1. (a) C-V profiles of $400^{\circ}C$-annealed sample at $\pm 3V$ to $\pm 8V$ gate voltage sweep. (b) A comparison of C-V profiles for as-deposited and $400^{\circ}C$-annealed samples.

Figure 2 presents the retention time characteristics of NFGM device containing the $400^{\circ}C$-annealed nanocomposite layer. At $\pm 5V$ voltage stress, the C-V curves were measured at different time intervals up to 10^4 sec. As shown in Fig. 2, there is a ΔV_{FB} shift about 3.50V and a charge loss about 28.4% after the retention time of 10^4 sec. This illustrates a satisfactory charge retention property for NFGM sample containing $AIST-SiO_2$ nanocomposite layer.

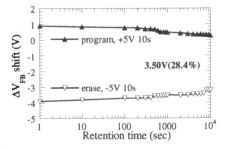

Figure 2. Retention time characteristics of NFGM device containing $400^{\circ}C$-annealed nanocomposite layer subjected to $\pm 5V$ voltage stress.

TEM characterization

Cross-sectional TEM (XTEM) image of $400^{\circ}C$-annealed $AIST-SiO_2$ nanocomposite layer is

185

shown in Fig. 3(a). EDX analysis indicated the atomic ratio of Ag:In:Sb:Te is about 0.3:6.1:63.2:30.4, implying the stoichiometric ratio of AIST is $Ag_{0.01}In_{0.2}Sb_2Te$ or the main chalcogenide phase type is Sb_2Te. Figure 3(b) shows an enlarged portion of Fig. 3(a) in which the lattice fringes in AIST nanoparticles delineate they are in crystalline form. This confirms previous argument that the nonvolatile memory characteristic of NFGM devices is resulted from the presence of metallic nanoparticles in nanocomposite layer. Furthermore, a SiO_x layer about 3 to 5 nm thick is observed in between the region enriched with AIST nanoparticles and Si substrate. It is speculated that such an interfacial oxide layer may serve as the tunneling layer of NFGM devices during the electrical property characterizations.

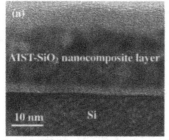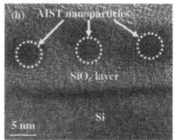

Figure 3. (a) XTEM image of 400°C-annealed AIST-SiO₂ nanocomposites layer and (b) enlarged micrograph of (a) at nanocomposite/Si substrate interface.

XPS analysis

Figure 4 displays the XPS spectra and their de-convoluted profiles obtained by Gaussian curve fitting method[19] for the samples prior and posterior to the 400°C-annealing. Sputtering etching removal of about 25-nm thick nanocomposites layer was performed prior to analysis so as to ensure the XPS data were extracted from the region enriched with AIST phase. As shown in Fig. 4(a), SiO_2 and suboxide bonds with a small amount of Si-Si bond are observed in as-deposited sample and reduction of Si-O bonds to form Si-Si bonds occurs after the annealing treatment. A much obvious oxide reduction to metallic element is deduced from Figs. 4(b) and 4(c) which illustrate nearly all antimony oxides, e.g., Sb_2O_5 and Sb_2O_3, in as-deposited sample transform to metallic Sb in annealed sample. As to Te element, Te $3d_{5/2}$ XPS depth profiles depicted in Fig. 4(d) indicate Te presents in oxide form (TeO_2) in the region near to Al electrode while in metallic form in the region enriched with AIST phase. Annealing treatment barely affects the chemical status of Te. It is speculated that such an oxide layer may serve as the capping layer and is able to reduce the leakage current of devices; however, accumulation of TeO_2 at nanocomposite/Al electrode interface and its inertness to heat treatment require further study. XPS analysis indicates the key process to improve the nonvolatile memory characteristics

186

of NFGM device is the reduction of antimony oxides to metallic Sb which, in turn, results in the increment of metallic nanoparticles in AIST-SiO$_2$ nanocomposite layer. Above characterizations also imply the NFGM device can be fabricated *via* a one-step sputtering process and the incorporation of AIST-SiO$_2$ nanocomposite layer yields a simplified device structure and processing method for NFGM device fabrication.

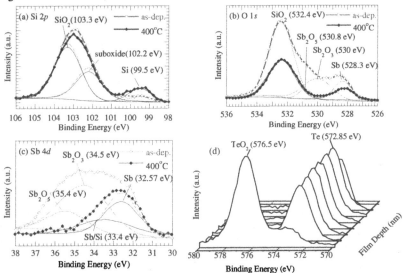

Figure 4. (a) Si 2p (b) O 1s (c) Sb 4d XPS spectra for 400°C-annealed and as-deposited AIST-SiO$_2$ nanocomposite layers. Raw XPS profiles are decorated with solid or open diamonds while the corresponding curve fitting results are represented by solid and grey curves. (d) Te 3$d_{5/2}$ XPS depth profiles of 400°C-annealed sample.

CONCLUSIONS

This study demonstrates the feasibility of AIST-SiO$_2$ nanocomposite layer to NFGM device fabrication *via* sputtering deposition. Significantly large ΔV_{FB} shift = 5.91V and charge density = 5.22×10^{12} cm^{-2} at ±8V voltage sweep were observed in the sample containing nanocomposite layer subjected to 400°C annealing for 2 min. The ΔV_{FB} shift about 3.50V and the charge loss about 28.4% were observed after the retention time of 10^4 sec at ±5V gate voltage stress. According to TEM and XPS characterizations, the improvement of charge-storage trap ability is ascribed to the reduction of antomony oxides to metallic Sb which results in the increment of metallic nanoparticles in AIST-SiO$_2$ nanocomposite layer subjected to appropriate annealing treatment.

ACKNOWLEDGMENTS

This work is supported by National Science Council (NSC), Taiwan, R.O.C., under the contract No. NSC97-2221-E-009-029-MY3.

REFERENCES

1. S. Tiwari, F. Rana, K. Chan, H. Hanafi, C. Wei, and D. Buchanan, Tech. Dig.-Int. Electron Devices Meet. **1995**, 521.
2. C. C. Wang, J. Y. Wu, Y. K. Chiou, C. H. Chang, and T. B. Wu, Appl. Phys. Lett. **91**, 202110 (2007).
3. P. Prakaipetch, Y. Uraoka, T. Fuyuki, A. Tomyo, E. Takahashi, T. Hayashi, A. Sano, and S.Hori, Appl. Phys. Lett. **89**, 093502 (2006).
4. D. W. Kim, T. Kim, and S. K. Banerjee, IEEE Trans. Electron Devices **50**, 1823 (2003).
5. P. F. Lee, X. B. Lu, J. Y. Dai, H. L. W. Chan, E. Jelenkovic, and K. Y. Tong, Nanotechnology, **17**, 1202, (2006).
6. S. W. Ryu, Y. K. Choi, C. B. Mo, S. H. Hong, P. K. Park, and S. W. Kang, J. Appl. Phys. **101**, 026109 (2007).
7. J. H. Kim, K. H. Baek, C. K. Kim, Y. B. Kim, and C. S. Yoon, Appl. Phys. Lett. **90**, 123118 (2007).
8. Y. S. Jang, J. H. Yoon, and R. G. Elliman, Appl. Phys. Lett. **92**, 253108 (2008).
9. Dufourcq, P. Mur, M. J. Gordon, S. Minoret, R. Coppard, and T. Baron, Mater. Sci. Eng., **C27**, 1496 (2007).
10. S. K. Samanta, W. J. Yoo, G. Samudra, E. S. Tok, L. K. Bera, and N. Balasubramanian, Appl. Phys. Lett. **87**, 113110 (2005).
11. Z. Liu, C. Lee, V. Naratanan, G. Pei, and E. C. Kan, IEEE Trans. Electron Devices **49**, 1606 (2002).
12. X. J. Wang, L. D. Zhang, M. Liu, J.P. Zhang, and G. He, Appl. Phys. Lett. **92**,122901 (2008).
13. C. C. Wang, J. Y. Tseng, T. B. Wu, L. J. Wu, C. S. Liang, and J. M. Wu, *Appl. Phys. Lett.* **99**, 026102 (2006).
14. C. C. Chou, F. Y. Hung, and T. S. Lui, Scripta Materialia **56**,1107 (2007).
15. D. R. Lide, CRC Handbook on Chemistry and Physics (A CRCnet BASE Product, Taylor and Francis Group, LLC), 89th ed., (2008-2009), 12-114.
16. H.-C. Mai and T.-E. Hsieh, Jpn. J. Appl. Phys. **46**, 5834 (2007).
17. H.-C. Mai, T.-E. Hsieh, S.-H. Huang, S.-S. Lin, and T.-S. Lee, Jpn. J. Appl. Phys. **47**(2008), 6029.
18. S. Maikap, S. Z. Rahaman, and T. C. Tien, Nanotechnology **19**, 435202 (2008).
19. J. F. Moulder, W. F Stickle, P. E. Sobol, and K. D. Bombem, Handbook of X-ray Photoelectron Spectroscopy, 2nd ed., Physical Electronics, Minnesota, 1992.

Mater. Res. Soc. Symp. Proc. Vol. 1250 © 2010 Materials Research Society   1250-G16-02

Effect of W substitution in Strontium Bismuth Tantalate Ferroelectric Ceramics: Enhanced Ferroelectric properties

Indrani Coondoo[1], Neeraj Panwar[2] A. M. Biradar[1] and A. K. Jha[3]

[1]National Physical Laboratory, Dr. K. S. Krishnan Road, New Delhi – 110012

[2]University of Puerto Rico, San Juan, Puerto Rico, USA, PR-00931

[3]Thin Film and Materials Science Laboratory, Department of Applied Physics, Delhi

Technological University, Delhi -110042

ABSTRACT

Tungsten (W)-substituted SBT ceramics [$SrBi_2(Ta_{1-x}W_x)_2O_9$; $0.0 \leq x \leq 0.20$] were synthesized by solid state reaction method using different sintering temperatures (1100 °C, 1150 °C, 1200 °C and 1250 °C). W substitution is found to significantly affect the electrical properties of SBT, including dielectric permittivity, Curie temperature, and ferroelectricity. Dielectric constant (ε_r) and the Curie temperature (T_c) increase with increasing W content. The dielectric loss reduces significantly with increase in W concentration. The maximum T_c of ~ 390 °C is observed in the sample with $x = 0.20$ as compared to ~ 320 °C for the pure sample when sintered at 1200 °C. The peak ε increases from ~ 270 in the sample with $x = 0.0$ to ~ 700 for the composition with $x = 0.20$, when sintered at 1200 °C. All the tungsten-substituted ceramics have higher $2P_r$ than that in the pristine sample. The maximum $2P_r$ (~25 $\mu C/cm^2$) is obtained in composition with $x = 0.05$ sintered at 1200 °C. These effects have been interpreted based on the model of the recovery of oxygen vacancies upon W substitution. Such compositions with low loss and high P_r values should be excellent materials for highly stable ferroelectric memory devices.

INTRODUCTION

Ferroelectrics are exceedingly useful materials in modern technology, with applications such as transducers, actuators, dielectrics, and nonvolatile memories. Among ferroelectrics, it was observed that bismuth oxide layered structures (e.g. $SrBi_2Ta_2O_9$, $BaBi_2Ta_2O_9$, $SrBi_2Nb_2O_9$) originally synthesized by Aurivillius are the most suitable ones for NvRAMs. Since Araujo et. al. [1] reported the fatigue free behavior of $SrBi_2Ta_2O_9$ (SBT), it has occupied an important position in Pb-free piezoelectrics as well as realization of ferroelectric nonvolatile memories (Fe-RAM).

The crystal structure of SBT consists of $(Bi_2O_2)^{2+}$ layers and perovskite-type $(SrTa_2O_7)^{2-}$ units with double TaO_6 octahedral layers [2,3]. One interesting feature of the Aurivillius phases resides in the compositional flexibility of the perovskite blocks which allows incorporating various cations for the A- and B-sites. It is thus possible to modify the ferroelectric properties according to the chemical composition. The ferroelectricity arises mainly in the perovskite blocks; the ferroelectricity is attributed to the rotation and tilting of TaO_6 octahedra as well as the off-center displacement of Ta ions in the octahedral unit in SBT [4,5].

Here we report the influence of tungsten (W) substitution and sintering temperature on the dielectric and ferroelectric properties of SBT.

EXPERIMENTAL DETAILS

Samples of series $SrBi_2(Ta_{1-x}W_x)_2O_9$ (SBTW), with x = 0.0, 0.025, 0.050, 0.075, 0.10 and 0.20 were synthesized by conventional solid-state reaction method using $SrCO_3$, Bi_2O_3, Ta_2O_5 and WO_3 (from Aldrich) in their stoichiometric ratios. The powder mixtures were thoroughly ground and passed through sieve of appropriate size and then calcined at 900 °C in air for 2 hours. The calcined mixtures were ground and admixed with about 1-1.5 wt % polyvinyl alcohol (Aldrich) as a binder and then pressed at ~300 MPa into disk shaped pellets. The pellets were air-sintered at 1100 °C, 1150 °C, 1200 °C and 1250 °C for 2 hours.

The sintered pellets were polished to a thickness of 1mm and coated with silver paste on both sides for use as electrodes and cured at 550 °C for half an hour. The dielectric measurements were carried out using a Solartron1260 Gain-phase analyzer operating at oscillation amplitude of 50mV. The Polarization-Electric field (P-E) hysteresis measurements were recorded at room temperature using an automatic PE loop tracer based on Sawyer -Tower circuit.

RESULTS AND DISCUSSION

Dielectric studies

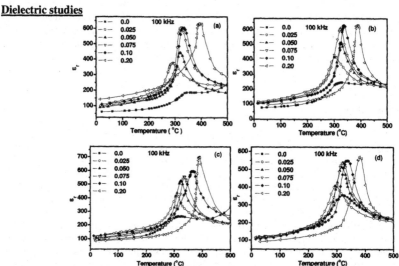

Figure 1 Variation of ε_r with temperature at 100 kHz in W- substituted samples sintered at (a) 1100°C, (b) 1150°C, (c) 1200°C and (d) 1250°C.

190

Figure 1 (a-d) shows the temperature dependence of ε_r (obtained at 100 khz) of the studied samples. All the SBTW samples exhibit sharp transition at their respective T_c while pure SBT shows broadened transition. In order to analyze the effect of sintering temperature on the peak-ε_r and T_c, figures 2(a and b) are plotted which exhibit the variation of peak-ε_r and T_c with sintering temperatures, respectively. A shift in T_c to higher temperatures and a corresponding increase in the peak dielectric constant values with increasing concentration of tungsten are observed. In all cases, T_c decreases for samples with $x = 0.025$ but shows an increasing trend over the composition range of $x = 0.05 - 0.20$. It also shows that the optimum sintering temperature for maximum dielectric constant and T_c is 1200 °C.

Generally in isotropic perovskite ferroelectrics, doping at B-site (located inside an oxygen octahedron) with smaller ions results in the shift of the Curie point to a higher temperature, leading to a larger polarization due to the enlarged "rattling space" available for smaller B-site ions [6]. However, in the anisotropic layered-perovskites, the crystal structure may not change as freely as that in the isotropic perovskites with doping due to the structural constraint imposed by the $(Bi_2O_2)^{2+}$ interlayer [7,8]. Moreover, since the valency of the substitutional cation (W^{6+}) is higher than the Ta^{5+}, substitution creates cationic vacancies at Sr-site ($V_{Sr}^{''}$) to maintain charge neutrality of the lattice structure [7-9]. The corresponding defect representation can be expressed as:

$$Null = W_{Ta}^{\bullet} + \frac{1}{2}V_{sr}^{''}$$

(1)

where $W_{Ta}^{\bullet}$ implies W occupying the Ta-site. When the W-content is low, the lattice structure under the constraint of Bi-O layer and the presence of cation vacancies at the A-site possibly have resulted in an increased stress value. In such a situation the perovskite structure would be

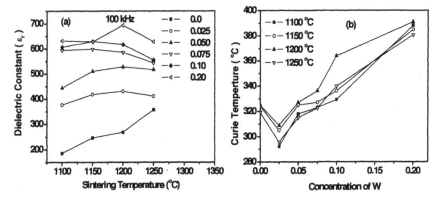

Figure 2 (a) Variation of peak dielectric constant with sintering temperature in $SrBi_2(Ta_{1-x}W_x)_2O_9$ samples observed at 100 kHz and (b) Variation of Curie temperature with W concentration at different sintering temperatures.

less stable and may cause a decrease in T_c [10] as observed for samples with $x = 0.025$ (figure 2a). On comparing the variation of lattice parameters a and b (table 1) with tungsten concentration, we observe that for tungsten concentrations $x > 0.05$, the decrease in the lattice parameters a and b is much greater than that observed for concentrations $x \leq 0.05$. This indicates that at higher concentrations the structural constraint imposed by the Bi-O interlayer was overcome resulting in structural distortion. It is this enhancement of ferroelectric structural distortion along with the introduction of cation vacancies at the A-site that lead to an eventual increase in Tc value [7, 11].

Table 1. Lattice parameters a, b and c of $SrBi_2(W_xTa_{1-x})_2O_9$ samples sintered at 1100°C, 1150°C, 1200°C 1250°C.

x	1100 °C			1150 °C			1200 °C			1250 °C		
	a (Å)	b (Å)	c (Å)	a (Å)	b (Å)	c (Å)	a (Å)	b (Å)	c (Å)	a (Å)	b (Å)	c (Å)
0	5.5285	5.5128	25.0131	5.5243	5.5337	25.1001	5.5212	5.5139	24.9223	5.5143	5.5246	24.9017
0.025	5.5103	5.5005	24.9882	5.5215	5.5090	24.9767	5.5314	5.5202	25.1079	5.5237	5.5323	25.0678
0.05	5.5102	5.4964	25.0253	5.5172	5.5062	25.0046	5.5270	5.5199	25.0585	5.5219	5.5361	25.0366
0.075	5.508	5.4985	25.0349	5.5005	5.4807	25.0156	5.5251	5.5045	25.0567	5.4854	5.5059	24.8736
0.1	5.5104	5.4979	25.0426	5.4822	5.4689	25.0354	5.5242	5.5060	25.085	5.4782	5.4972	25.0867
0.2	5.502	5.4817	25.0603	5.4852	5.4628	25.0701	5.5233	5.4939	25.0861	5.4786	5.4955	25.1515

Also, the high T_c is indicative of enhanced polarizability [7], that explains the increase in peak-ε_r with increase in tungsten concentration. Moreover, it has been reported that the cation vacancies make the domain motion easier and increases the dielectric permittivity [12,13] resulting an increase in ε_r with increasing W content. The densification and microstructural development due to the compositional deviation from the stoichiometry affect the dielectric properties, as well. The peak-ε_r value of the pristine SBT increases with sintering temperature. However, in the SBTW compositions, peak-ε_r value increases with the sintering temperature up to 1200 °C and show a decrease when sintered at 1250 °C (figure 2a). The above trends in ε_r for both pure and SBTW ceramics are quite similar to that of densification [14], indicating that densification has an effect on ε_r.

Ferroelectric studies

Figure 3 shows P-E loops (for samples with $x = 0.025$ and 0.05) and the compositional dependence of remnant polarization (P_r) of SBTW samples prepared at different sintering temperatures. As can be observed, P_r depends on W concentration as well sintering temperature.

192

For the same sintering temperature $2P_r$ first increases with x and then decreases. The optimum tungsten content for maximum $2P_r$ (~ 25 $\mu C/cm^2$) is observed to be $x = 0.05$. It is also noticeable that for the same tungsten content, there is an optimum sintering temperature for maximum $2P_r$ (1200 °C for $x = 0.025 - 0.1$ and 1250 °C for $x = 0.0$). The $2P_r$ value decreases from ~ 25 $\mu C/cm^2$ for composition with x = 0.075 sintered at 1200 °C to ~ 17 $\mu C/cm^2$ when sintered at 1250 °C.

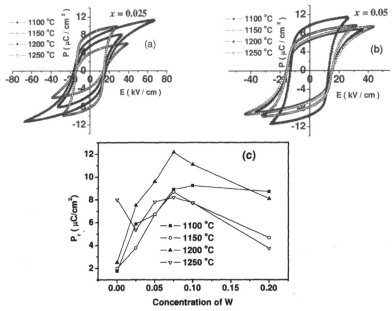

Figure 3   P-E loops for $x =$ (a) 0.025 and (b) 0.05 prepared at different sintering temperatures. (c) Compositional variation of $2P_r$ of SBTW samples prepared at different sintering temperatures (1100, 1150, 1200 and 1250 °C).

It is known that ferroelectric properties are affected by compositional modification microstructure and lattice defects like intrinsic oxygen vacancies [15,16]. In soft ferroelectrics with higher-valent substituents, the defects are cation vacancies whose mobility is extremely low below T_c. Thus the interaction between cation vacancies and domain walls is much weaker than that in hard ferroelectrics (with lower-valent substituents), wherein the associated mobile oxide vacancies are likely to assemble in the vicinity of domain walls locking them and making their polarization switching difficult, leading to a decrease in P_r values [17]. Based on the obtained results and above discussion, it can be understood that in pure SBT, the oxygen vacancies assemble at sites like domain boundaries leading to strong domain pinning. Whereas in the SBTW samples, the associated cation vacancy formation due to the substitution of Ta^{5+} by W^{6+} suppresses the concentration of oxygen vacancies. A reduction in the number of oxygen vacancies reduces the pinning effect on the domain walls, leading to enhanced remnant polarization. Also, it is known that domain walls are relatively free in large grains and are inhibited in their movement as the grain size decreases [18]. In the larger grains, domain motion is easier which results in higher P_r values [19]. In the present study, the grain size is observed to increase with increasing W concentration (not shown here); however, the remnant polarization does not monotonously increase with increasing W concentration (Figure 3). The increase in the number of charge carriers in the form of oxygen vacancies beyond $x > 0.05$ leads to pinning of domain walls and thus a reduction in the values of P_r is observed.

CONCLUSIONS

In the present study, it can be concluded that the solubility limit of tungsten content in SBT is $x \leq 0.05$ and their electrical properties reveals an optimum sintering temperature of 1200 °C. Tungsten doping is effective in enhancing the dielectric and ferroelectric properties. Dielectric constant and T_c increases with increasing W content. All the tungsten-substituted ceramics have higher $2P_r$ than that in pure samples. The maximum $2P_r$ (~ 25 $\mu C/cm^2$) is obtained in the composition with $x = 0.05$ sintered at 1200°C. Such compositions with low loss and high P_r values should be excellent materials for highly stable ferroelectric memory devices.

REFERENCES

1. C. A. P. de Araujo, J. D. Cuchiaro, L. D. McMillan, M. C. Scott, and J. F. Scott, *Nature* **374**, 627 (1995).
2. E. C. Subbarao, *Integr. Ferr.* **12**, 33 (1996).
3. Y. Wu, C. Nguyen, S. Seraji, M. J. Forbess, S. J. Limmer, T. Chou and G. Z. Cao, *J. Amer. Ceram. Soc.* **84**, 2882 (2001).
4. P. D. Martin, A. Castro, P. Millan and B. Jimenez, *J. Mater. Res.* **13**, 2565 (1998).
5. Y. Noguchi, M. Miyayama and T. Kudo, *Phys. Rev. B* **63**, 214102 (2001).
6. I. Coondoo, A K Jha and S K Agarwal, *J. Eur. Ceram. Soc.* **27**, 253 (2007).
7. H. T. Martirena and J. C. Burfoot, *J. Phys. C: Solid State Phys.* **7**, 3182 (1974).
8. Y. Noguchi, M. Miyayama and T. Kudo, *J. Appl. Phys.* **88**, 2146 (2000).
9. K. Singh, D. K. Bopardik and D. V. Atkare, *Ferroelectrics* **82**, 55 (1988).
10. H. Watanabe, T. Mihara, H. Yoshimori and C. A. P. De Araujo, *Jpn. J. Appl. Phys.* **34**, 5240 (1995).
11. T. Atsuki, N. Soyama, T. Yonezawa and K. Ogi, *Jpn. J. Appl. Phys.* **34**, 5096 (1995).
12. T. Friessnegg, S. Aggarwal, R. Ramesh, B. Nielsen, E. H. Poindexter and D. H. Keeble, *Appl. Phys. Lett.* **77**, 127 (2000).
13. Y. H. Xu, *Ferroelectric Materials* (Elsevier Science Publishers, Amsterdam, 1991).
14. I. Coondoo, N. Panwar and A. K. Jha, *J. Am. Ceram. Soc.* (Communicated, 2010).
15. H. Watanabe, T. Mihara, H. Yoshimori and C. A. P. De Araujo, *Jpn. J. Appl. Phys.* 34, 5240 (1995).
16. M. Miyayama, T. Nagamoto and O. Omoto, *Thin Sol. Films* **300**, 299 (1997).
17. Y. H. Xu, *Ferroelectric Materials* (Elsevier Science Publishers, Amsterdam, 1991)
18. R. R. Das, P. Bhattacharya, W. Perez and R. S. Katiyar, *Ceram. Int.* **30**, 1175 (2004).
19. S. B. Desu, P. C. Joshi, X. Zhang and S. O. Ryu, *Appl. Phys. Lett.* **71**, 1041 (1997).

Mater. Res. Soc. Symp. Proc. Vol. 1250 © 2010 Materials Research Society   1250-G17-03

Characteristics of Organic Memory Using Metal Oxide Nano-Clusters

You-Wei Cheng[1], Tzu-Yueh Chang[1,2], and Po-Tsung Lee[1,2]
[1]Department of Photonics & Display Institute, National Chiao Tung University, 1001 University Road, Hsinchu 300, Taiwan
[2]Department of Photonics & Institute of Electro-Optical Engineering, National Chiao Tung University, 1001 University Road, Hsinchu 300, Taiwan

ABSTRACT

In this report, electrical properties of an organic memory device with a tri-layer structure, MoO_3 nano-clusters layer sandwiched between Alq_3 thin films, are investigated. The device using this kind of structure exhibits a large ON/OFF current density ratio over 10^4, long retention time over 1hr, and an electrically programmable character. The formation of the bistable resistance switching of the device originates from a charge trapping effect of the MoO_3 nano-clusters layer. Moreover, the current density-voltage (J-V) characteristics of the device are quite different from those of organic bistable devices (OBDs) using MoO_3 nano-particles. No negative differential resistance is observed in the J-V curve of the device. This may be due to the distinct surface morphology of the MoO_3 layer on the Alq_3 thin film.

INTRODUCTION

In past decades, conjugated organic materials have been widely applied in organic electronic and optoelectronic devices such as organic thin film transistors, organic light emitting diodes, organic photovoltaic cells, etc. One of candidates for next generation memory devices, an organic memory device, is emerging because of greater scope for better scalability, low-cost fabrication, mass production capability, and mechanical flexibility. Many published results have paid attention to organic memories with nano-structured materials [1-7] (e.g., nano-cluster/organic-metal composite layers [1-3], nano-composites of polymer-gold nano-particles [4-6]) inside an organic layer(s) as charge trapping centers. Because of distinct properties of the nano-structured materials, organic resistance switching memories with high density, large ON/OFF ratio, and other superior performances can be obtained in the near future. Molybdenum trioxide (MoO_3) has been extensively applied in organic optoelectronics (e.g., as a doping layer to raise conductivity [8]). Such wide applications are attributed to the characteristics of the MoO_3 thin film: high work function, high conductivity for holes, and high transparency. In this report, electrical characteristics of an OBD with an Alq_3/MoO_3 nano-clusters /Alq_3 tri-layer structure are studied. The OBD shows a large ON/OFF current density ratio over 10^4, long retention time over 1hr, and a rewritable/reerasable feature. The resistance switching mechanism of the OBD resulted from a charge trapping effect of the MoO_3 nano-clusters layer. Moreover, no negative differential resistance (NDR) showed in the J-V curves of OBDs using MoO_3 nano-particles [7] is observed in the J-V curve of the OBD. This may ascribe to the different surface morphology of the MoO_3 layer on the Alq_3 thin film. Besides, the simple structure of the OBD indicates that it can be easily embedded into the well-developed semiconductor fabrication processes.

EXPERIMENT

The OBD consisted of an Alq_3/MoO_3 nano-clusters/Alq_3 tri-layer structure interposed between anode and cathode in this study is shown in Fig. 1(a). First, a p^+-type silicon substrate was cleaned according to a standard RCA clean process. A 50 nm thick Alq_3 thin film was evaporated onto the cleaned p^+-type silicon substrate below 3×10^{-6} Torr at room temperature. The structural formula of Alq_3 is shown in Fig. 1(b). Then, 5 nm thick MoO_3, and 50 nm thick Alq_3 thin films were evaporated in sequence onto the Alq_3/p^+-Si. The average deposition rate of the Alq_3 thin films and that of the MoO_3 layer were about 0.1 nm/s and 0.01 nm/s, respectively. Finally, a 100 nm thick Al thin film was evaporated through metal mask with 2 mm x 2 mm square patterns onto Alq_3/MoO_3 nano-clusters/Alq_3/p^+-Si as top electrode. The J-V characteristics of the OBD were measured using a Hewlett Packard 4156A semiconductor parameter analyzer in ambient environment. The p^+-type silicon substrate was kept at 0 V, and all bias conditions were applied on the aluminum electrode. The surface morphology of the MoO_3 layer on the Alq_3 thin film was detected using an atomic force microscope (DI-Veeco Instrument).

(a)   (b)

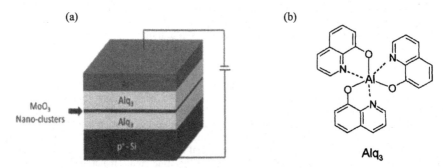

Figure 1. (a) Schematic diagram of an OBD using an Al/Alq_3/MoO_3 nano-clusters/Alq_3/p^+-Si structure. (b) Structural formula of Alq_3.

DISCUSSION

The J-V characteristics of an OBD using an Al/Alq_3/MoO_3 nano-clusters/Alq_3/p^+-Si structure are shown in Fig. 2(a). At first sweeping (the red curve), the sweeping bias from 0 to 10 V is applied. Initially, the OBD is in a low conductance state. An abrupt increase in the current density level is observed when the applied voltage is over the threshold voltage of the OBD. Then, the OBD holds at a high conductance state when the applied voltage sweeps to higher voltage. In other words, an electrical transition from an OFF state to an ON state is observed in the OBD. At following sweeping, the OBD still maintained at the high conductance state (the blue curve). It is obvious that there is a clear difference between two conductance states with an ON/OFF current density ratio over 10^4. It suggests that the OBD displays the nature of bistability. In addition, a striking decrease in the current density level is observed when the OBD is applied a reversed sweeping bias form 0 to -10 V (the green curve). The OBD is switched from the high

(a)

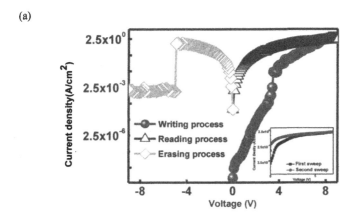

(b)

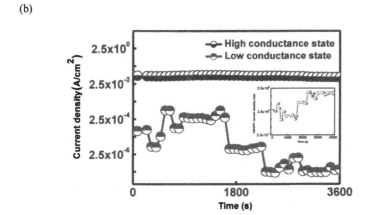

Figure 2. (a) J-V characteristics of an OBD with an Al/Alq₃/MoO₃ nano-clusters/Alq₃/p⁺-Si structure. The red circles, blue triangles, and green diamonds represent the writing, reading, and erasing sweeping biases, respectively. Inset: The J-V curves of Ag/Alq₃/P⁺-Si. (b) Retention measurement of the OBD. The blue and red circles correspond high and low conductance states, respectively. Inset: ON/OFF current density ratio as a function of time.

conductance state to the low conductance state. The OBD could be switched to the high conductance state again when a following sweeping bias from 0 to 10 V is applied. A sweeping bias form 0 to 10 V and a reversed polarity sweeping bias form 0 to -10 V can be treated as"writing" and "erasing" processes of the OBD, respectively. Thus, the OBD holds two main characteristics of a memory: bistability and rewritability.

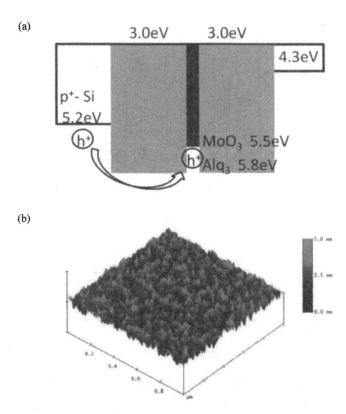

Figure 3. (a) Energy band diagram of the Al/Alq₃/MoO₃/Alq₃/p⁺-Si structure. (b) 3-D surface morphology of the 5 nm thick MoO₃ layer deposited on the Alq₃ /p⁺-Si.

The nature of bistability is ascribed to holes trapped in the MoO₃ nano-clusters layer due to the confinement of energy level difference between Alq₃ and nano-structured MoO₃ as shown in Fig. 3(a). No switching behavior is observed in the J-V curves of a device without MoO₃ nano-clusters layer (Ag/Alq₃/p⁺-Si), as shown in the inset of Fig. 2(a). In addition, NDR is found in the J-V characteristics of OBDs using a MoO₃ nano-particles layer [7], but it is not observed in Fig. 2(a). Such discrepancy possibly results from the different surface morphologies of the MoO₃ layers. As shown in Fig. 3(b), the surface morphology of a 5 nm thick MoO₃ layer deposited on the Alq₃/p⁺-Si exhibits a nano-cluster feature instead of a nano-particle feature. The space charge field resulted from charge trapping of the MoO₃ nano-clusters layer is more random than that of the MoO₃ nano-particles layer. This means the effective space charge field effect on the carrier transportation of the MoO₃ nano-clusters layer is weaker than that of the MoO₃ nano-particles layer. Consequently, after the OBD is switched from the low conductance state to

198

the high conductance state, the OBD could maintain at the high conductance state without NDR. We can conclude that the surface morphology of a nano-structured MoO_3 layer plays an important role in the electrical characteristics of OBDs using nano-structured MoO_3.

Retention time is one important property of the OBD. The retention measurement of the OBD is measured by applying a voltage bias at 1 V and the result is shown in Fig. 2(b). There is no recognizable change in the current density of the high conductance state. On the contrary, current density perturbation is observed in the low conductance state. The current density perturbation of the low conductance state results from the incomplete erasing of trapped charges after the OBD is applied a reversed sweeping bias. However, the current density of the low conductance state is getting smaller and more stable with time, and the ON/OFF current density ratio becomes larger. An evident conductance difference between ON and OFF states could be distinguished, and the device possessed long retention time over 1hr.

CONCLUSION

The electrical characteristics of an OBD using a MoO_3 nano-clusters layer are reported. The bistability of the OBD is a consequence of the charge trapping effect of the MoO_3 nano-clusters layer interposed between Alq_3 thin films. No NDR is observed in the J-V characteristics of the OBD because of a weaker effective space charge field resulted from the nano-cluster feature of the MoO_3 layer. Thus, the surface morphology of a nano-structured MoO_3 layer makes a great influence on the electrical characteristics of OBDs using a nano-structured MoO_3 layer.

REFERENCES

1. L. P. Ma, J. Liu, and Y. Yang, "Organic electrical bistable devices and rewritable memory cells," Appl. Phys. Lett., vol. 80, pp. 2997-2999 (2002).
2. L. D. Bozano, B. W. Kean, V. R. Deline, J. R. Salem, and J. C. Scott, "Mechanism for bistability in organic memory elements," Appl. Phys. Lett., vol. 84, pp. 607-609 (2004).
3. L. D. Bozano, B. W. Kean, M. Beinhoff, K. R. Carter, P. M. Rice, and J. C. Scott, "Organic materials and thin-film structures for cross-point memory cells based on trapping in metallic nanoparticles," Adv. Funct. Mater., vol. 15, pp. 1933-1939 (2005).
4. J. Ouyang, C.-W. Chu, C. R. Szmanda, L. Ma, and Y. Yang, "Programmable polymer thin film and non-volatile memory device," Nature Mater., vol. 3, pp. 918-922 (2004).
5. J. Ouyang, C.-W. Chu, D. Sieves, and Y. Yang, "Electric-field-induced charge transfer between gold nanoparticle and capping 2-naphthalenethiol and organic memory cells," Appl. Phys. Lett., vol. 86, pp. 123507 (2005).
6. R. J. Tseng, J. Huang, J. Ouyang, R. B. Kaner, and Y. Yang, "Polyaniline nanofiber/gold nanoparticle nonvolatile memory," Nano Lett., vol. 5, pp. 1077-1080 (2005).
7. K. S. Yook, S. O. Jeon, C. W. Joo, J. Y. Lee, S. H. Kim, and J. Jang, "Organic bistable memory device using MoO_3 nanocrystal as a charge trapping center," Org. Electron., vol. 10, pp. 48-52 (2009).
8. G. Xie, Y. Meng, F. Wu, C. Tao, D. Zhang, M. Liu, Q. Xue, W. Chen, and Y. Zhao, "Very low turn-on voltage and high brightness tris-(8-hydroxyquinoline) aluminum-based organic light-emitting diodes with a MoO_x p-doping layer," Appl. Phys. Lett., vol. 92, pp. 093305 (2008).

Resistance RAM (ReRAM)

Mater. Res. Soc. Symp. Proc. Vol. 1250 © 2010 Materials Research Society   1250-G05-03

Size-dependent temperature instability in NiO–based resistive switching memory

D. Ielmini, F. Nardi, C. Cagli and A. L. Lacaita
Dipartimento di Elettronica e Informazione – Politecnico di Milano, Italy
Phone: +30 02 2399 6120, email: ielmini@elet.polimi.it

ABSTRACT

Resistive switching memory (RRAM) is attracting a strong interest as novel nonvolatile memories for high-density storage. Anyway this technology has to overcome two main issues before its use in real applications which are the high current needed for program operations and data retention stability. These two problems are here investigated from experimental and theoretical points of view to clarify the possibilities of NiO RRAMs to become a real competitive alternative to mainstream Flash technology.

INTRODUCTION

Resistive-switching RAM (RRAM) is a new kind of nonvolatile memory characterized by the ability to reversibly change the resistance of an active layer, typically a transition metal oxide, under appropriate electrical control. Several materials, e.g. NiO [1], TiO_2 [2], Cu_2O [3], display this ability using both bipolar and unipolar programming voltages. Resistive switching in NiO can be unipolar [1], where resistance change is due the formation and dissolution of a conductive filament (CF) across the oxide layer, or bipolar [4], due to field-driven migration of oxygen vacancies. In unipolar switching the CF is dissolved (reset operation) through Joule heating and subsequent thermal oxidation [5, 6], while it is created (set operation) through threshold switching effect [6, 7], where a purely electrical enhancement of conductivity generates a high localized temperature leading to a local chemical reduction of the oxide phase to the metallic one [8, 9]. Thanks to its unipolar resistive-switching, NiO-based RRAMs are very interesting for high-density arrays beyond the 10-nm technology node, with the implementation in a diode-selected crossbar architecture. The main problem for this kind of applications is the high current (I_{reset}) needed for reset operation, usually above 100 μA [10]. This may limit the dimensions of the select diode, hence the minimum size of the memory array. To reduce I_{reset}, a proper control of the CF size is necessary, e.g. limiting the set current [11].

Here we present the impact of CF size on unipolar NiO RRAMs reliability. We show that the decrease of CF size, i.e. increase of resistance, impact negatively data retention and positively reset current. These results can be interpreted by a size-dependent diffusion-oxidation model of the CF.

FILAMENT NATURE AND SIZE ESTIMATION

Experiments were performed on NiO-based RRAM cells fabricated by MDM Laboratories [12]. The simple metal-insulator-metal (MIM) structure of these cells is composed as follows: a W bottom electrode with diameter from 0.18 to 1 μm, a NiO film deposited by atomic layer deposition (ALD) with a thickness $t_{NiO} = 20$ nm, and a top Pt electrode. The first programming operation on the cell is forming, consisting of a dielectric breakdown with the first formation of a

CF through the oxide layer. Then the cell can be electrically switched between set state (low resistance, i.e. 10^2 Ω) and reset state (high resistance, i.e. 10^8 Ω).

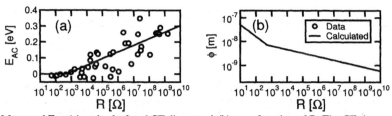

Fig. 1: Measured E_{AC} (a) and calculated CF diameter ϕ (b), as a function of R. The CF size was obtained by the piecewise linear fit of E_{AC} in (a), where $E_{AC} = 0$ for metallic-like and $E_{AC} > 0$ for semiconductor-like behaviors.

Important hints about the physical nature of the CF have been obtained by selecting different values of resistance R using the incremental reset algorithm [13] and observing the resistance behavior at variable temperature T. A metallic behavior for $R < 0.5$ kΩ was found, where the resistance increases with T according to:

$$R = R_0[1 + \alpha(T - T_0)] \tag{1}$$

where R_0 is the resistance at the reference temperature, e.g. room temperature T_0, and α is the temperature coefficient for resistance [5]. Instead, for $R > 0.5$ kΩ, the resistance decreases with T, according to the Arrhenius law:

$$R = \rho_0 \frac{t_{NIO}}{A} e^{\frac{E_{AC}}{kT}} \tag{2}$$

where ρ_0 is the Arrhenius pre-exponential factor for resistivity, A is an effective area for the CF and E_{AC} is the activation energy for conduction. For these values of R the Arrhenius trend suggests a semiconductor nature of the CF. It is noteworthy that in this formula we considered the dependence of R from temperature and CF geometry but neglected the Meyer-Neldel effect on the pre-exponential factor ρ_0 [14]. In Fig. 1a are reported E_{AC} values, extracted from Arrhenius plots of the resistance, for different programmed R. Despite the significant spread of data, the increase of E_{AC} with R is clear. The metallic behavior for low R values suggests the presence of a CF with a fully metallic phase, e.g. Ni, while the semiconductor behavior for high R values suggests a suboxide composition of the conductive path, i.e. NiO_x with x <1, where point defects, e.g. oxygen vacancies, may contribute to thermally activated hopping conduction mechanism [7]. In fact, states with relatively high R were obtained by the partial reset operation, consisting of a partial oxidation of the full set-state (metallic filament) [13]. For increasing reset voltage, hence increasing oxidation of the CF, the resistance increased toward the pristine NiO value. It is thus reasonable to describe the intermediate state as NiO_x (x < 1), i.e. an intermediate composition between the metallic filament (Ni) and the quasi-stoichiometric NiO composition.

From the conductive properties of the CF, an estimate of the CF diameter ϕ can be obtained. Previous studies have attempted the direct characterization of CF size by CAFM techniques [15] or cross-sectional TEM [16], however an indirect estimation of ϕ or of the cross section A =

$\pi \phi^2/4$ from electrical data might be extremely useful for testing and characterization purposes [17]. The calculated ϕ as a function of R is shown in Fig. 1b. For the calculation, E_{AC} was assumed to obey the piecewise-linear fitting in Fig. 1a. It is shown that a change of more than 7 decades in R is reflected by a change in ϕ by a mere factor 20. This emphasizes the importance of including the E_{AC} dependence in Eq. (2) for CF size estimation. In fact, such a big variation of R cannot be explained by geometrical considerations only, rather the exponential dependence of E_{AC} must be taken into account.

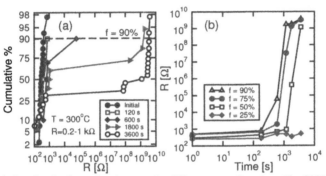

Fig. 2: Cumulative distributions of resistances for different annealing time (T = 300°C) (a) and constant-f data-loss characteristics (b).

DATA RETENTION

Decreasing the size of the CF is of critical importance for reducing I_{reset} [10], however small CFs may be affected by relatively fast data loss [18,19]. In fact the CF dissolution process is related to a T-activated diffusion and oxidation of metallic species, which can result in data loss at relatively low T [6,10]. To study the possible resistance dependence of data retention, we performed T-accelerated bake experiments in an annealing chamber with air ambient conditions. The annealing times were sufficiently long (at least 2 minutes) that the transient heating and cooling times of the devices under test could be considered negligible. Before each annealing experiment RRAM cells were programmed at room temperature not only in the lower R state but also in intermediate states. Then the cells were subjected to annealing with T ranging from 240 to 350 °C and finally their resistance was measured at room temperature. All the bake experiments were conducted on a statistically-relevant set of 50 cells in order to average the generally large spread of retention behaviors. Fig. 2a shows the cumulative distributions of R for increasing annealing times at 300°C, for cells with initial R between 0.2 and 1 kΩ. R increases for increasing times due to the oxidation of the metallic CF. Starting from the cumulative distributions in Fig. 2a it is possible to extract the time evolution of R at constant percentile f. This constant-f data-loss characteristic is reported in Fig. 2b and can be considered as the typical retention behavior of memory cells around that percentile [20].

Fig. 3a shows the Arrhenius plot for the retention time τ_R, here defined as the time for a 10x increase of R in the constant-f data-loss characteristics. τ_R is reported for increasing percentiles, from $f = 25\%$ to 90%, and for an initial resistance between 0.2 and 1 kΩ. Less data points are shown for increasing f because these display increasingly longer τ_R, that could not always be detected within the measurement time (see e.g. data for $f = 25\%$ in Fig. 2b). Data show an Arrhenius dependence with an activation energy around $E_{AR} = 1.2$ at $f = 50\%$. To clarify possible correlations of retention time with CF size, Fig. 3b shows the Arrhenius plot of τ_R, for $f = 90\%$ and for three different ranges of R, i.e. R < 0.2 kΩ, 0.2 < R < 1 kΩ, and R > 1 kΩ. From these data we can clearly see that τ_R decreases for increasing R, i.e decreasing ϕ. Considering other values of f, not reported here, a similar relationship between τ_R and R is expected as can be inferred by the similar behaviors for different f reported in Fig. 3a.

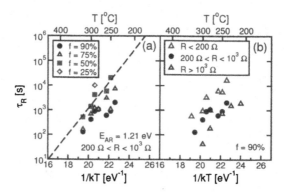

Fig. 3 Arrhenius plot of the measured τ_R values for different f (25%, 50%, 75% and 90%) for R ranging from 200 Ω to 1 kΩ (a), and Arrhenius plot of measured τ_R for three different resistance ranges (R < 0.2 kΩ, 0.2 < R < 1 kΩ, and R > 1 kΩ) at $f = 90\%$ (b). The x-axis reports $1/kT$, where T is the temperature in the annealing experiments. Resistance was instead measured at room temperature.

DATA RETENTION MODEL

To explain the dependence of CF size with data retention, we considered the oxidation of the CF to be limited by the diffusion of metallic atoms, i.e. Ni, [21]. Dissolution of the CF might in fact be due to diffusion of nickel atoms from the Ni-rich CF toward the Ni-poor, O-rich surrounding phase, where Ni is oxidized. This diffusion process was simulated using a 3D finite-difference model. Fig. 4a shows the results of these simulation where the values of retention time τ_R for different CF sizes are reported. In this simulation τ_R was considered as the time for a decrease of the metallic concentration in the CF by a factor of 10 or 100. These calculated τ_R show a direct proportionality with ϕ^2. Considering both the temperature dependence (Arrhenius model) and the ϕ^2 geometry dependence of τ_R we can write:

$$\tau_R = \tau_{R0} \left(\frac{\phi}{\phi_0} \right)^2 e^{\frac{E_{AR}}{kT}}$$

(3)

where τ_{R0} and ϕ_0 are constants. Note that the Meyer-Neldel effect on τ_{R0} was not considered for simplicity [22]. Interpolating data in Fig. 3b, for a fixed temperature T, we can obtain the

206

experimental τ_R which are reported in Fig. 4b for T = 250°C. Calculations are obtained by Eq. (3), using extracted ϕ from Fig. 1b and parameter values E_{AR} = 1 eV, τ_{R0} = 1.4x10^{-7} s and ϕ_0 = 5 nm. The good agreement between data and calculations supports the validity of our physical interpretation for data loss mechanism. Extrapolations at 85°C are also shown in comparison with the 10 years criterion generally used for reliability assessment in non-volatile memories. From these results we can see that the CF size should be maximized, i.e. R minimized, for best data retention.

To highlight the tradeoff between retention time and reset current, Fig. 4c shows measured I_{reset} values as a function of R and calculations from the formula:

$$I_{reset} = (R^{-1}_{th}R^{-1}\Delta T_{reset})^{1/2} \qquad (4)$$

where R_{th} is the effective thermal resistance and ΔT is the temperature increase $T_{reset} - T_0$ in which T_0 is room temperature and T_{reset} is the temperature necessary for reset operation obtained by Eq. (3) for a specific τ_R [17]. It is clear that lower R show a longer τ_R but a higher I_{reset} [10]. This fundamental tradeoff between data retention and reset current can only be solved by accurate material engineering, e.g. increasing the activation energy of the solid-state oxidation to improve data retention time at smaller reset current.

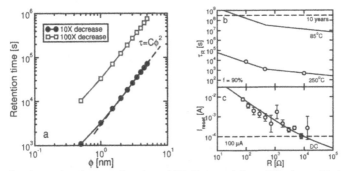

Fig. 4 Calculated retention time as a function of CF diameter ϕ from diffusion-oxidation model (a). Measured and calculated τ_R (b) and DC measured and calculated I_{reset} (c) as a function of R. Experimental data loss values are taken from interpolations in Fig. 3b at T=250°C. Calculations are shown for T = 250°C (same as reported data) and 85°C for reliability assessment of RRAM.

CONCLUSIONS

This work addresses the reliability of unipolar RRAMs based on NiO, as a function of the size of the metallic CF. The retention time τ_R was found to decrease for increasing resistance, i.e. decreasing CF size. These data are well interpreted by a diffusion-oxidation model. From these results the tradeoff between data retention and reset current is addressed both from experimental and theoretical points of view. These findings contribute to assess the feasibility and reliability of NiO RRAMs as next high-density nonvolatile memory solution.

ACKNOWLEDGMENTS

The authors would like to thank S. Spiga, E. Cianci and M. Fanciulli of MDM Laboratories for providing experimental samples and for insightful comments.

REFERENCES

1. I. G. Baek, M. S. Lee, S. Seo, M. J. Lee, D. H. Seo, D.-S. Suh, J. C. Park, S. O. Park, H. S. Kim, I. K. Yoo, U.-In Chung and J. T. Moon, IEDM Tech. Dig. 587-590 (2004).
2. D. Strukov, G. S. Snider, D. R. Stewart and R. S. Williams, Nature 453, 80-83 (2008).
3. A. Chen, S. Haddad, Y.-C. Wu, T.-N. Fang, L. Zhida, S. Avanzino, S. Pangrle, M. Buynoski, M. Rathor, W. Cai, N. Tripsas, C. Bill, M. VanBuskirk, M. Taguchi, IEDM Tech. Dig. 746-749 (2005).
4. L. Goux, J. G. Lisoni, L. Courtade, C. Muller, M. Jurczak and D. J. Wouters, IEEE Proceedings of International Memory Workshop, 13 (2009).
5. U. Russo, D. Ielmini, C. Cagli and A. L. Lacaita, IEEE Trans. Electron Devices 56, 186-192 (2009).
6. C. Cagli, F. Nardi and D. Ielmini, IEEE Trans. Electron Devices 56, 1712-1720 (2009).
7. D. Ielmini, Phys. Rev. B 78, 035308 (2008)
8. D. Ielmini, C. Cagli and F. Nardi, Appl. Phys. Lett. 94, 063511 (2009)
9. C. H. Kim, H. B. Moon, S. S. Min, Y. H. Jang and J. H. Cho, Solid-State Comm. 149, 1611-1615 (2009).
10. U. Russo, D. Ielmini, C. Cagli and A. L. Lacaita, IEEE Trans. Electron Devices 56, 193-200 (2009).
11. K. Kinoshita, K. Tsunoda, Y. Sato, H. Noshiro, S. Yagaki, M. Aoki and Y. Sugiyama, Appl. Phys. Lett. 93, 033506 (2008).
12. S. Spiga, A. Lamperti, C. Wiemer, M. Perego, E. Cianci, G. Tallarida, H. L. Lu, M. Alia, F. G. Volpe, M. Fanciulli, , Microelectronic Engineering 85, 2414-2419 (2008).
13. D. Ielmini, F. Nardi, A. Vigani, E. Cianci and S. Spiga, IEEE Semiconductor Interface Specialist Conference (SISC), Arlington, VA, Dec. 3-5, 2009.
14. R. S. Crandall, Phys. Rev. B 43, 4057 (1991).
15. T. Ohgai, L. Gravier, X. Hoffer, M. Lindeberg, K. Hjort, R. Spohr and J.-Ph. Ansermet, J. Phys. D: Appl. Phys. 36 3109-3114 (2003).
16. N. D. Davydov, J. Haruyama, D. Routkevitch, B. W. Statt, D. Ellis, M. Moskovits and J. M. Xu, Phys. Rev. B 57, 13550-13553 (1998).
17. D. Ielmini, F. Nardi, C. Cagli and A. L. Lacaita, IEEE Electron Device Lett., in press
18. T.-N. Fang, S. Kaza, S. Haddad, A. Chen, Y.-C. Wu, Z. Lan, S. Avanzino, D. Liao, C. Gopalan, S. Choi, S. Mahdavi, M. Buynoski, Y. Lin, C. Marrian, C. Bill, M. VanBuskirk, and M. Taguchi, in IEDM Tech. Dig., 2006, pp. 789–792.
19. J. Park, J. Yoon, M. Jo, D.-J. Seong, J. Lee,W. Lee, J. Shin, E.-M. Bourin, and H. Hwang, in as discussed at 2009 IEEE SISC, Arlington, VA, 2009
20. D. Ielmini, A. S. Spinelli, A. L. Lacaita and A. Modelli, IEEE Electron Device Lett. 23, 40-42 (2002).
21. S. A. Makhlouf, Thin Solid Films 516, 3112 (2008).
22. D. Ielmini and M. Boniardi, Appl. Phys. Lett. 94, 091906 (2009).

Mater. Res. Soc. Symp. Proc. Vol. 1250 © 2010 Materials Research Society   1250-G05-05

Correlation between Oxygen Composition and Electrical Properties in NiO Thin Films for Resistive Random Access Memory

Yusuke Nishi, Tatsuya Iwata, and Tsunenobu Kimoto
Department of Electronic Science and Engineering, Kyoto University, Kyotodaigaku-katsura, Nishikyo, Kyoto, 615-8510, Japan

ABSTRACT

Admittance spectroscopy measurement has been performed on NiO_x thin films with various oxygen compositions (x=1.0-1.2) in order to characterize localized defect levels. The activation energy and concentration of localized defect levels in NiO_x films with low oxygen composition ($x \leq 1.07$) are 120-170 meV and lower than 2×10^{19} cm^{-3}, respectively. From I-V measurement of $Pt/NiO_x/Pt$ structures, samples with high oxygen composition ($x \geq 1.10$) did not show resistance switching operation, while samples with low oxygen composition ($x \leq 1.07$) did. The best oxygen composition of NiO_x thin films turned out to be 1.07 in order to realize repeatable and stable resistance switching operation.

INTRODUCTION

In recent years, a variety of nonvolatile memories have emerged and been developed as the next-generation nonvolatile random access memories (RAMs), which may replace the Flash memories. Resistive RAMs (ReRAMs) using binary transition metal oxides such as NiO [1,2], CoO [3], Cu_xO [4] or TiO_2 [5,6] have attracted extensive interest owing to many advantages of low-power, high-speed operation, high on/off ratio and compatibility with complementary metal-oxide-semiconductor (CMOS) technologies. Despite its promising properties, the lack of fundamental understanding of resistance switching mechanism in metal oxide thin films has hindered the industrial application of ReRAMs.

In the present study, admittance spectroscopy measurement has been performed on NiO_x thin films with various oxygen compositions deposited on n-Si substrates. The oxygen composition x was intentionally changed in the wide range from 1.0 to 1.2. Current-voltage (I-V) measurement of $Pt/NiO_x/Pt$ structures on p-Si substrates has also been conducted. The correlation between the oxygen composition and these electrical properties is discussed.

EXPERIMENT

NiO thin films were deposited on n-Si substrates by a reactive radio-frequency (RF) sputtering method using an Ni target of 99.99% (4N) purity. The ratio of O_2 flow rate in the Ar + O_2 gas mixture was varied in the range from 3% to 10% to change the oxygen composition. The total pressure, substrate temperature and RF power during sputtering were kept at 1.0 Pa, 300°C and 100W, respectively. The thickness of NiO films was 200 to 300 nm. From the cross-sectional high-resolution transmission electron microscopy (HRTEM), the deposited films are a mixture of both polycrystalline and amorphous regions. The oxygen composition x of NiO_x films was determined as 1.0 to 1.2 (metal deficient p-type semiconductor [7]) by using Rutherford backscattering (RBS) and energy dispersive X-ray spectroscopy (EDX). Pt electrodes with a typically 300 μm diameter were deposited by electron beam evaporation through a metal mask.

Admittance spectroscopy measurement has been performed on NiO_x thin films in order to characterize localized defect levels. Samples with Pt/p-NiO_x/n-Si/Al structure were fabricated. The donor concentration of n-Si substrates was in the 10^{19} cm^{-3} range. The frequency dependence of conductance and capacitance was measured in vacuum using the Yokogawa Hewlett Packard 4192A LF Impedance Analyzer. The temperature range for the measurement was from 150K to 400K.

Current-voltage (I-V) measurement of Pt/NiO_x/Pt structures on p-Si substrates has also been conducted in vacuum using the Keithley 4200 Semiconductor Parameter Analyzer. During I-V measurements, the bottom electrode was grounded and the bias voltage was applied to the top electrode. The temperature dependence of I-V characteristics was carried out in the temperature range from 300K to 550K.

RESULTS AND DISCUSSION

Localized states characterization in NiOₓ by admittance spectroscopy

Admittance spectroscopy is useful for characterization of localized defect levels in semiconductor bulk or films. The conductance G and the capacitance C of a depletion layer in a semiconductor region are given by the following equation [8]:

$$G = \frac{\omega^2 \tau}{1 + \omega^2 \tau^2} \Delta C \,, \tag{1}$$

$$C = C_{HF} + \frac{\Delta C}{1 + \omega^2 \tau^2} \,, \tag{2}$$

with

$$\Delta C = C_{LF} - C_{HF} \,, \tag{3}$$

where ω is the probe angular frequency, τ is emission time constant. C_{LF} and C_{HF} are the low- and high-frequency capacitances, respectively. Figure 1 shows ideal frequency dependence of G/ω (if G_{dc} is negligible) and C in the material with a single defect level. As predicted by equations (1) and (2), G/ω shows a single peak and C declines dramatically near the point at $\omega\tau = 1$. The emission time constant τ is given by

$$\tau(T) = \frac{1}{N_v(T)\langle v_{th}(T)\rangle_p \sigma_p(T)} \exp\left(-\frac{\Delta E}{kT}\right). \tag{4}$$

Here N_v is the effective density of states in the valence band, $\langle v_{th}\rangle$ is the mean thermal velocity of holes, σ_p is the hole capture cross section, ΔE is the activation energy of the hole traps, respectively.

Figure 2 shows the capacitance-voltage (C-V) characteristics of the typical p-$NiO_{1.07}$/n-Si heterojunction. The diffusion potential and the net acceptor concentration of $NiO_{1.07}$ were estimated to be 0.8 eV and in the 10^{18} cm^{-3} range, respectively. The depletion layer width of this junction was about 10 nm at zero bias, which is much thinner than the thickness of NiO films.

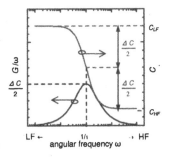

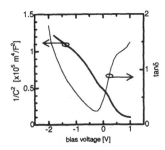

Figure 1. Ideal frequency dependence of G/ω and C in the material with a single localized level.

Figure 2. Bias voltage dependence of $1/C^2$ and tanδ in the p-NiO$_{1.07}$/n-Si heterojunction.

Figure 3 shows the plots of $(G-G_{dc})/\omega$ and C measured at zero bias for the p-NiO$_{1.07}$/n-Si heterojunction as a function of frequency in the temperature range from 250K to 360K. Here G is the conductance of the investigated samples, and G_{dc} is the DC component of the conductance, which can be approximated to be the conductance at 10 Hz. Note that, the sharp rise of $(G-G_{dc})/\omega$ at about 10 MHz was confirmed to originate from the resonance caused by the measured capacitance and parasitic inductance of the measurement system. The result reveals that the NiO$_{1.07}$ thin film has a single defect level, of which emission time constant decreases with increase in temperature.

Figure 4 shows the plots of $(G-G_{dc})/\omega$ and C measured at zero bias and at room temperature (RT) for the p-NiO$_x$/n-Si heterojunctions with various oxygen compositions as a function of frequency. The localized levels in NiO$_x$ thin films strongly depend on the oxygen composition. For the NiO$_x$ films with low oxygen composition ($x \leq 1.07$), only a single peak is observed in the $(G-G_{dc})/\omega$ curve, while two peaks (or one peak and a shoulder) exist for the NiO$_x$ films with higher oxygen composition ($x \geq 1.15$). For the high composition films, the capacitance values exhibit steady rise toward the low-frequency region, indicating an additional level. All of the peak frequency decrease with increase in temperature.

The Arrhenius plots of τT^2 obtained from admittance spectroscopy on various NiO$_x$ films are shown in Fig. 5; when two defect levels exist in an NiO$_x$ film, the second level which has longer emission time constant than the first one is shown in gray color. Here τ is multiplied by T^2, taking the temperature dependence of N_v and $\langle v_{th} \rangle$ into account. The energy level of observed defects located above the valence band edge (equivalent to ΔE) in each NiO$_x$ film is added in the legend of Fig. 5. The activation energy of defects in NiO$_x$ films with low oxygen composition ($x \leq 1.07$) is relatively large, 120-170 meV, while that with higher oxygen composition ($x \geq 1.10$) is smaller than 100 meV. Furthermore, N_t/N_s and N_s can be estimated from the frequency dependence of $(G-G_{dc})/\omega$ and C, and the C-V measurement at high frequency where the holes trapped at defect levels can not follow, respectively. Here N_t is the concentration of hole trap levels, N_s the concentration of shallow acceptors. Table 1 shows N_s and N_t in the NiO$_x$ with various oxygen compositions at room temperature (RT). N_t is higher than N_s, which suggests that most holes in NiO$_x$ films are trapped.

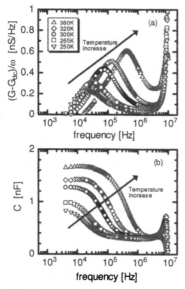

Figure 3. Frequency dependence of (a) $(G\text{-}G_{dc})$ /ω and (b) C for the p-NiO$_{1.07}$/n-Si heterojunction at different temperature.

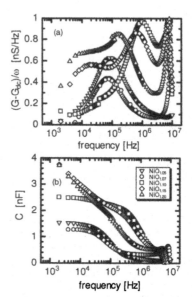

Figure 4. Frequency dependence of (a) $(G\text{-}G_{dc})$ /ω and (b) C for various p-NiO$_x$/n-Si heterojunctions at room temperature.

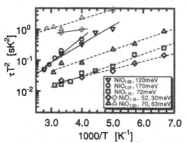

Figure 5. Arrhenius plot of τT^2 obtained from Admittance spectroscopy. The calculated energy levels of defects located above the valence band edge are added in a legend.

Table 1. Properties at room temperature of the peaks obtained from Admittance spectroscopy on the NiO$_x$ thin films with various oxygen compositions.

x in NiO$_x$	N_t [cm⁴]	τ [μs]	ΔE [meV]	N_t [cm⁴]
1.05	2×10^{18}	1.8	120	1×10^{19}
1.07	5×10^{18}	2.3	170	1×10^{19}
1.10	8×10^{18}	0.18	72	5×10^{19}
1.15	1×10^{19}	0.24	52	5×10^{19}
		8.8	30	8×10^{18}
1.20	1×10^{19}	0.88	70	5×10^{19}
		11	63	1×10^{19}

Resistance switching characteristics of Pt/NiO$_x$/Pt structures

Figure 6 shows the oxygen composition dependence of initial resistance (R$_{ini}$) in the Pt/NiO$_x$/Pt structures with a top-electrode (TE) diameter of 300 μm at RT. Values of resistance were measured at 0.3 V. The current compliance was set to 10 mA in order to avoid damage. R$_{ini}$ exhibits remarkable decrease with increase in the oxygen composition x of NiO$_x$ thin films. Samples with high oxygen composition (x≥1.10) did not show resistance switching operation because R$_{ini}$ is very low and the current reached the compliance level of even above 10 mA before the electrical "forming" process occurred, while samples with low oxygen composition (x≤1.07) did; especially in the Pt/NiO$_{1.07}$/Pt structures it was repeatable and stable.

Here, we focus on the resistance switching characteristics of the Pt/NiO$_{1.07}$/Pt structure. The resistance switching characteristics at various temperature are shown in Fig 7. The on-off ratio, which is the ratio of the resistance in high-resistance state (R$_{HRS}$) to that in low-resistance state (R$_{LRS}$), is about 800 at 300 K and about 40 at 550 K. These ratios are sufficiently large to read dynamic margin for practical switching operation. Moreover, R$_{ini}$ and R$_{HRS}$ exhibited very similar temperature dependence (not shown). The Arrhenius plots of R$_{ini}$ and R$_{HRS}$ result in the activation energy of 170 meV; it should be noted that this value is similar to ΔE (170 meV) estimated above from the emission time constant in the p-NiO$_{1.07}$/n-Si heterojunction.

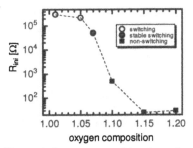

Figure 6. Oxygen composition dependence of initial resistance (R$_{ini}$) in the Pt/NiO$_x$/Pt structures with a top-electrode diameter of 300 μm at room temperature.

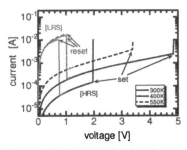

Figure 7. Resistance switching characteristics of the Pt/NiO$_{1.07}$/Pt structure with a top-electrode diameter of 300 μm at 300 K, 400 K and 550 K.

Correlation between Oxygen Composition and Electrical Properties

Samples with high oxygen composition (x≥1.10) did not show resistance switching operation due to the low initial resistance. On the other hand, samples with low oxygen composition (x≤1.07) did; especially in the Pt/NiO$_{1.07}$/Pt structures it was repeatable and stable. These different properties of stable resistance switching may originate from the energy level of the hole trap located above the valence band edge.

Figure 5 indicates that the emission time constant (τ) of hole trap levels tends to decrease with increase in oxygen composition of NiO$_x$. According to this tendency, the additional peak

213

with much smaller emission time constant than the observed peak (τ=2.3 μs at RT) may exist in the Pt/NiO$_{1.07}$/Pt structure which shows repeatable and stable resistance switching operation. The desirable range of oxygen composition of NiO$_x$ thin films is narrow in order to realize repeatable and stable resistance switching operation.

CONCLUSIONS

Admittance spectroscopy measurement has been performed on NiO$_x$ thin films with various oxygen compositions (x=1.0-1.2). Several properties of hole trap levels in NiO$_x$ thin films have been characterized. The activation energy of localized states in NiO$_x$ films with low oxygen composition ($x \leq 1.07$) are relatively large, 120-170 meV, while that with higher oxygen composition ($x \geq 1.10$) is smaller than 100 meV. The concentration of localized states in NiO$_x$ films is lower than 2×10^{19} cm^{-3}. The best oxygen composition of NiO$_x$ thin films turned out to be about 1.07 in order to realize repeatable and stable resistance switching operation. Since the present Pt/NiO$_{1.07}$/Pt structure can exhibit repeatable resistance switching operation even at temperature as high as 550 K, the NiO-based ReRAM has shown promise for high-temperature applications in the future.

REFERENCES

1. J. F. Gibbons and W. E. Beadle, *Solid State Electron.*, **7**, 785 (1964).
2. I. G. Baek, M. S. Lee, S. Seo, M. J. Lee, D. H. Seo, D.-S. Suh, J. C. Park, S. O. Park, H. S. Kim, I. K. Yoo, U-In Chung, and J. T. Moon, *Tech. Digests of the 2004 IEEE Int. Electron Devices Meet.*, pp. 587-590.
3. H. Shima, F. Takano, H. Akinaga, Y. Tamai, I. H. Inoue, and H. Takagi, *Appl. Phys. Lett.* **91**, 012901 (2007).
4. A. Chen, S. Haddad, Y.-C. Wu, T.-N. Fang, Z. Lan, S. Avanzino, S. Pangrle, M. Buynoski, M. Rathor, W. Cai, N. Tripsas, C. Bill, M. VanBuskirk, and M. Taguchi, *Tech. Digests of the 2005 IEEE Int. Electron Devices Meet.*, pp. 746-749.
5. F. Argall, *Solid State Electron.* **11**, 535 (1968).
6. B. J. Choi, D. S. Jeong, S. K. Kim, C. Rohde, S. Choi, J. H. Oh, H. J. Kim, C. S. Hwang, K. Szot, R. Waser, B. Reichenberg, and S. Tiedke, *J. Appl. Phys.*, **98**, 033715 (2005).
7. H. J. Van Daal and A. J. Bosman, *Phys. Rev.*, **158**, 736 (1967).
8. J. L. Pautrat, B. Katircioglu, N. Magnea, D. Bensahel, J. C. Pfister and L. Revoil, *Solid State Electron.*, **23**, 1159 (1980).

Mater. Res. Soc. Symp. Proc. Vol. 1250 © 2010 Materials Research Society   1250-G05-06

Effect of Ion-implantation on Forming and Resistive-Switching Response of NiO Thin films

Robert G. Elliman, Muhammad N. Saleh, Sung Kim, Dinesh K. Venkatachalam, Taehyun Kim and Kidane Belay

Electronic Materials Engineering Department, Research School of Physics and Engineering, Australian National University, Canberra ACT 0200, Australia

ABSTRACT

The forming voltage and set/reset response of sputter-deposited NiO thin films is studied as a function of implant fluence for samples implanted with Ni and O ions. The forming voltage of the films is shown to decrease with increasing ion fluence and to scale with the damage production rate of the different ions. In contrast, the set/reset response of the films was largely unaffected by the ion-implantation. These results are discussed in terms of the filamentary model of conduction and the thermochemical model of resistive switching.

INTRODUCTION

Resistive switching is a process in which the resistance of a dielectric thin film is repeatedly switched between low and high resistance states by the application of appropriate current-voltage pulses, and is believed to result from the local breaking and reforming of conductive filaments produced by the initial voltage stress (forming process). This process was first reported more than fifty years ago [1-3] but has recently received renewed attention because of its potential for fabricating non-volatile memory devices [4-5]. However, in order to fully exploit it the atomic mechanisms responsible for the switching behavior must be understood. Many models for have been proposed, including trap charging in the dielectric, space-charge-limited conduction processes, ion conduction and electrodeposition, Mott transition, and Joule-heating effects. In some cases, such as TiO_2, the understanding is now highly developed [6] while in others the understanding remains rudimentary and/or speculative.

NiO is particularly attractive for non-volatile memory applications due to its high-speed, low power switching characteristics and its CMOS compatibility [7-9]. Resistive switching in this case is thought to be based on a thermochemical process in which a conductive filament formed by field-induced defects is broken (reset) by defect annealing caused by local Joule heating [10-12]. The electrical properties of NiO are a sensitive function of the film stoichiometry, and O-vacancies are known to play a significant role in determining such properties. The presence of metallic impurities, such as Ti or Li, is also known to affect the switching characteristics [13-14].

Ion-implantation can be used to controllably introduce defects and/or impurities into different regions of thin film structures and therefore provides a useful tool for studying the influence of such parameters on resistive switching. In this study we examine the effect of Ni and O ion-implantation on the forming voltage and set/reset response of NiO thin films.

EXPERIMENTAL

Metal-insulator-metal test structures were fabricated on oxidized (300nm thermal oxide) (100) Si wafers. The bottom contact was produced by sputter coating the wafer with a 15nm Ti wetting layer and a 200nm thick Pt contact layer. A 270nm thick NiO layer was subsequently deposited by reactive sputter deposition of Ni in a 10% O_2/ 90% Ar ambient. The substrate was held at room-temperature during deposition and the operating pressure was maintained at 4 mTorr. The final sample structure is shown schematically in figure 1a.

Portions of the as-deposited NiO film were subsequently irradiated at room-temperature with 270 keV Ni$^+$, 80 keV O$^-$, or both. The energy of the ions was chosen to give approximately the same mean projected range (~100 nm) and depth distribution for O and Ni. The resulting damage distribution extended to around 140 nm, approximately half of the NiO film thickness, as shown in figures 1b, c. Ion fluences were in the range from 5×10^{14} cm^{-2} to 5×10^{15} cm^{-2} and samples implanted with both ions received equal fluences of Ni$^+$ and O$^-$.

The physical structure of the as-deposited and ion-implanted films was studied by Rutherford backscattering spectrometry (RBS), glancing-incidence x-ray diffraction (GI-XRD) and transmission electron microscopy (TEM). Electrical characterization was performed using an Agilent B1500A Semicondutor Device Analyser. Samples were directly contacted on the top surface using a W probe and forming and set/reset response measured using a DC voltage ramp. A current compliance of 20 mA was used for forming and set measurements.

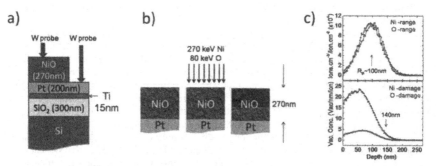

Figure 1: a) Schematic of sample test structure; b) Schematic of ion-implantation process, and; c) Ion-range and the vacancy distributions calculated by the SRIM code using the Kinchin-Pease estimate.

RESULTS AND DISCUSSION

As-deposited films

GI-XRD and TEM analysis of as-deposited NiO films showed that they were polycrystalline, with grains having a NaCl crystal structure, as shown in figures 2a, b. Filament forming was achieved by increasing the voltage across the 270nm film until dielectric breakdown was observed. The current was limited to 20 mA during forming to avoid irreversible damage to the layer. An average forming voltage of 20.5V was measured for the as-deposited films, which corresponds to an electric field of 760 kV/cm. After forming the films showed unipolar resistive switching characteristics, with typical set/reset behavior as shown in figure 2c. The average set voltage was around 1.5V (with a current compliance of 20 mA) and the average reset voltage was around 0.5V (with a typical reset current of around 35 mA).

Figure 2: a) GI-XRD spectrum of as-deposited NiO film; b) plan-view bright-field image of NiO film and associated diffraction pattern, and; c) typical set/reset response of the as-deposited film.

Ion-implanted films

As discussed above, and shown in figure 1c, the radiation damage created by O and Ni implantation extends to around 140 nm, approximately half of the NiO film thickness. It is also evident from figure 1c that Ni ions create more damage than O ions, with the integrated damage being 5 times greater for Ni ions than for O ions.

Despite the relatively high implant fluences employed in the current study, NiO films remain polycrystalline after irradiation, as shown in figure 3a. However, the diffraction peaks in GI-XRD spectra show broadening consistent with the presence of radiation induced defects in the film, as shown in figure 3b. The presence of these defects was found to have a direct effect on the forming voltage of test devices, with the average forming voltage decreasing with increasing ion-fluence. This is illustrated in figure 3c for O-implanted samples, where the forming voltage decreases from around 20 V for the as-deposited film to around 14 V for films irradiated to a fluence of 5.0×10^{15} O.cm^{-2}.

The effect of ion-implantation on the forming voltage is summarized in figure 4a for samples irradiated with O, Ni and O+Ni. In all cases the forming voltage is observed to decrease with increasing implant fluence, with the effect saturating at fluences greater than 5×10^{15} cm^{-2} in the case of O-implants, and for fluences greater than about 1×10^{15} cm^{-2} in the case of Ni-implants. The curves are well fitted by a simple exponential function of the form:

$$V_F = V_o + ae^{(-R\varphi)}$$

in which R parameterises the rate of change in the forming voltage, V_F, with ion fluence, φ. Comparison of the R values for the O and Ni data shows that they vary in the ratio 1:5, similar to the ratio of the damage produced by these ions. This suggests that the reduction in V_F is a direct consequence of defect production in the NiO layer rather than non-stoichiometry caused by the implanted O or Ni. The response of samples implanted with both Ni and O supports this view.

217

It is also interesting to note that the saturation value of V_F at high Ni fluences is around 10 V, a value approximately half that measured for the un-implanted film. This is consistent with the upper half of the film being rendered conductive by the implantation-induced defects.

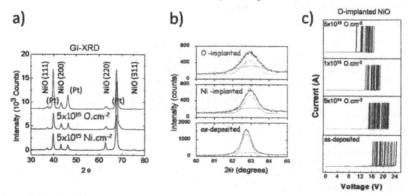

Figure 3: a) GI-XRD spectra of the as-deposited and ion-implanted NiO film; b) details of the NiO (220) reflection showing broadening after ion-implantation (Note: the shift in the peaks is instrumental error), and; c) Forming measurements for the as-deposited and O-implanted NiO film.

The effect of ion-implantation on the set/reset response of NiO films is summarised in figure 4b. The reliability of the switching response is reduced following high-fluence implantation but the average set and reset voltages remain unchanged. This is consistent with a thermochemical reset mechanism in which local Joule heating anneals defects within a local volume as such heating would also anneal implantation-induced defects thereby removing any 'memory' of the as-implanted defect state.

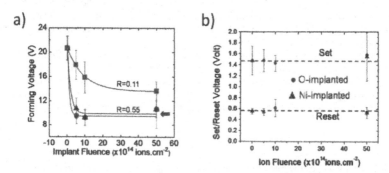

Figure 4: a) Forming voltage as a function of ion-fluence for NiO films implanted with O (■), Ni (▲) and O+Ni ions (●), and; b) Set/Reset voltages as a function of ion-fluence for NiO films implanted with O (●) or Ni (▲).

218

SUMMARY AND CONCLUSIONS

The forming voltage and set/reset response of NiO thin films was studied as a function of implant fluence for samples irradiated with Ni and O ions. The forming voltage was found to decrease with increasing ion fluence, with the magnitude of the effect scaling with the damage production rate of the different ions. For samples irradiated with high Ni fluences the forming voltage was reduced to approximately half of its initial value suggesting that the implanted upper half of the film was effectively conductive. As the defects created by the ions was limited to this region of the film this suggests that the implanted defect concentration exceeding the percolation threshold in the implanted region of the film. The reduction in forming voltage is consistent with the development of a defect mediated conductive filament in which defects produced by ion-implantation play a similar role to those produced by the applied field. In contrast, ion-implantation had little effect on the set/reset response of NiO films, consistent with thermochemical annealing of residual defects at the switching site.

ACKNOWLEDGMENTS

The authors acknowledge Mr David Llewellyn for assistance with the TEM sample preparation, and the Australian Research Council for partial financial support through the ARC Linkage Project scheme.

REFERENCES

1.   G. Dearnaley, A.M. Stoneham, and D.V. Morgan, *Rep. Prog. Phys.* **33**, 1129 (1970).
2.   J.F. Gibbons, and W.E. Beadle, *Solid-State Electron.* **7**, 785 (1964).
3.   T.W. Hickmott, *J. Appl. Phys.* **33**, 2669 (1962).
4.   R. Waser, *Microelectron. Eng.* **86**, 1925 (2009).
5.   D.B. Strukov, G.S. Snider, D.R. Stewart, and R.S. Williams, *Nature* **453**, 80 (2008).
6.   D.H. Kwon, K.M. Kim, J.H. Jang, J.M. Jeon, M.H. Lee, G.H. Kim, X.S. Li, G.S. Park, B. Lee, S. Han, M. Kim, and C.S. Hwang, *Nat. Nanotechnol.* **5**, 148 (2010).
7.   C. Kugeler, R. Weng, H. Schroeder, R. Symanczyk, P. Majewski, K.D. Ufert, R. Waser, and M. Kund, *Thin Solid Films* **518**, 2258 (2010).
8.   A. Demolliens, C. Muller, D. Deleruyelle, S. Spiga, E. Cianci, M. Fanciulli, F. Nardi, C. Cagli, and D. Ielmini, *Reliability of NiO-based resistive switching memory (ReRAM) elements with pillar W bottom electrode* (Proceedings of the IEEE International Memory Workshop, Monterey, CA, 2009) pp. 25-27.
9.   K. Tsunoda, K. Kinoshita, H. Noshiro, Y. Yamazaki, T. Iizuka, Y. Ito, A. Takahashi, A. Okano, Y. Sato, T. Fukano, M. Aoki, and Y. Sugiyama, *Low power and high speed switching of Ti-doped NiOReRAM under the unipolar voltage source of less than 3 V*, (Proceedings of the IEEE International Electron Devices Meeting, Washington, DC, 2007)pp. 767-770.
10.   U. Russo, D. Ielmini, C. Cagli, and A.L. Lacaita, *IEEE Trans. Electron Devices* **56**, 193 (2009).
11.   U. Russo, D. Ielmini, C. Cagli, and A.L. Lacaita, *IEEE Trans. Electron Devices* **56**, 186 (2009).

12. C. Cagli, F. Nardi, and D. Ielmini, *IEEE Trans. Electron Devices* **56**, 1712 (2009).
13. M.J. Lee, Y. Park, S.E. Ahn, B.S. Kang, C.B. Lee, K.H. Kim, W.X. Xianyu, I.K. Yoo, J.H. Lee, S.J. Chung, Y.H. Kim, C.S. Lee, K.N. Choi, and K.S. Chung, *J. Appl. Phys.* **103**, 4 (2008).
14. K. Jung, J. Choi, Y. Kim, H. Im, S. Seo, R. Jung, D. Kim, J.S. Kim, B.H. Park, and J.P. Hong, *J. Appl. Phys.* **103**, 4 (2008).

Mater. Res. Soc. Symp. Proc. Vol. 1250 © 2010 Materials Research Society 1250-G10-05

Memory Effect in Simple Cu Nanogap Junction

Hiroshi Suga[1], Masayo Horikawa[1], Hisao Miyazaki[2], Shunsuke Odaka[2], Kazuhito Tsukagoshi[2], Tetsuo Shimizu[1], and Yasuhisa Naitoh[1]
[1]Nanosystem Research Institute, National Institute of Advanced Industrial Science and Technology (AIST), 1-1-1, Higashi, Tsukuba, Ibaraki, Japan
[2] Research Center for Materials Nanoarchitectonics (MANA), National Institute for Materials Science (NIMS), 1-1 Namiki, Tsukuba, Ibaraki, Japan

ABSTRACT

We have investigated the resistance switching effect in Cu nanogap junction. Nanogap structures were created by means of electromigration and their electrical properties were measured in a high vacuum chamber. The measured current-voltage characteristics exhibited a clear negative resistance and memory effect with a large on-off ratio of over 10^5. The estimation from I-V curves indicates that the resistance switching was caused by the gap size change, which implies that the nanogap switching (NGS) effect also occurs in Cu electrodes, a popular wiring material in an integrated circuit.

INTRODUCTION

Resistive switching behavior of nanogap junction, which consists of two electrodes with a separation of about ten nanometer or below, is a very attractive subject of scientific and technical research.[1] It was found that the nanogap junction exhibits a reversible resistance switching effect that is dependent on the intensity of the applied voltage between the nanogap junction. We call the switching effect the "nanogap switch" (NGS) effect. This NGS device is nonvolatile, and has the potential for application in future nonvolatile memory devices.[2-4] It was suggested that the nonvolatile resistance changes in the NGS occur as a result of tunneling changes in gap width between the electrodes, which are caused by the applied bias voltage. Moreover, a clear negative differential resistance (NDR) is also shown in an I-V curve for all nanogap junctions that exhibit resistance switching.

The NGS device has a wide selectivity for electrode materials such as various metals[2], silicon[4], and graphene[5]. Since copper is the most popular material for wiring material in the integration circuit [6], it is expected that the NGS device with Cu electrodes is very advantages for industrial applications. On the other hand, when the line and space between Cu wires is achieved at single nanoscale level in near future, a leakage may be caused by the NGS effect. Therefore, it is important to evaluate its nanoscale behavior, such as that in a resistance switch in Cu nanogap junctions. It is anticipated that Cu electrodes in Cu nanogap junctions will readily become coated with layers of insulative oxide. Therefore, the effect without oxide layers on the NGS effect should be investigated.

In this letter, we investigated NGS effect in Cu nanogap junctions. In order to protect Cu surfaces form oxidization, the nanogap formations and electrical measurements were also carried out in the same vacuum chamber without exposing the samples to ambient air.

EXPERIMENT

Figure 1 shows diagram of sample fabrication. The nanogap junction was fabricated on a Si substrate coated with a 300 nm thick thermally oxidized layer. Nanowires with line width of 320 nm were patterned on the Si substrate coated with a 500 nm thick triple resist layer (Microchem, MMA/MAA copolymer, 495PMMA, and 950PMMA), using high-resolution electron beam lithography (Elionix, ELS7500). Cr (4 nm thick) and Cu (30 nm thick) layers were deposited on the substrate. A lift-off process was then performed in acetone, which resulted in patterned Cu nanowires. Nanogaps were fabricated on the Cu nanowire by controlled electromigration method using a vacuum probe station with a base pressure of approximately 1.8×10^{-4} Pa at room temperature. [7-10] The controlled electromigration method enabled to control of the gap width to less than several nanometers. After the fabrication, the electrical properties of the nanogap junctions were measured in the same probe station chamber using a semiconductor parameter analyzer (Keithley, 4200SCS).

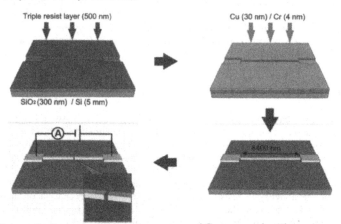

Figure 1. Schematic diagrams of fabrication process of Cu nanogap junction

RESULTS AND DISCUSSION

The typical I-V curves of a Cu nanogap junction are shown in Fig. 2(a). Since the I-V curves are symmetric, only the positive bias voltage region is shown. Here open squares are the results obtained when the voltage was varied from 0 to 10 V, and the open circles are those obtained when the voltage was varied from 10 to 0 V. As shown by open circles, the current increased as the voltage was decreased from 10 to 4.2 V, indicating a reducing of resistance. The current reduce as the voltage was decreased from 4.2 to 0 V, which indicates typical tunneling conduction. [11] The change in resistance indicates a typical negative differential resistance (NDR) behavior which was measured for all of the fabricated Cu nanogap junction over 1000 cycles.

Figure 2(b) shows typical resistance–voltage curves for Cu nanogap junction immediately after the voltage is changed from 10 to 0 V at a very fast rate of 5 µs/V, after which the voltage was slowly raised to 10 V at 0.167 s /V (open circles), and then returned from 10 to 0 V at the same rate of 0.167 s /V (open triangles). The open triangles clearly show the NDR behavior with a peak voltage of 4.2 V. Moreover, a clear resistance hysteresis was also observed between a very rapid lowering of the voltage and a slow lowering of the voltage below a bias voltage of 4.0 V. The resistance ratio measured at voltages lower than 0.5 V is over 10^4. A similar resistance hysteresis was observed in all of the fabricated Cu nanogap junctions. Here, the high resistance state when the voltage is rapidly reduced is called the "off-state", and the low resistance state when the voltage is slowly decreased is referred to as the "on-state".

In order to discuss the detail of the resistance changes, we estimated nanogap widths and other parameters by fitting experimental data to tunneling equation,[11] in which the tunneling current is expressed as

$$ I = \frac{k_1 A}{s^2} [x^2 \exp(-k_2 sx) - y^2 \exp(-k_2 sy)], \quad (1) $$

where $x = \sqrt{\varphi - V/2}$, $y = \sqrt{\varphi + V/2}$, $k_1 = 6.32 \times 10^{10}$ V s^{-1} and $k_2 = 1.025$ J$^{-1/2}$, and s, A, and φ represent the gap width, tunneling emission area, and the barrier height, respectively. Figure 3(a) and (b) show the experimental and fitting I-V curves of on and off state, respectively. The estimations were carried out within the range from -1 to +1 V. The results of the fitting were described in Table I. The gap widths of the on- and off-state are estimated to be 0.61 and 1.02 nm, respectively. The calculated gap size is in close agreement with gap size reported in Si nanogap junction.[5]

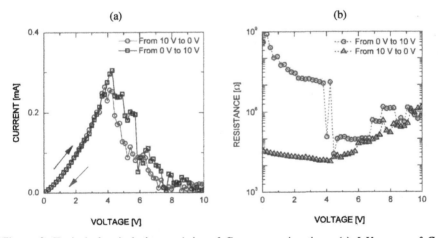

Figure 2. Typical electrical characteristics of Cu nanogap junctions. (a) I-V curves of Cu nanogap junction: open squares, from 0 to 10 V; open circles, from 10 to 0 V. (b) Resistive hysteresis between the curve representing resistance when the bias across the nanogap was increased from 0 to 10 V at 0.167 s /V, just after being lowered from 10 to 0 V at 5 μs/V (open circles) and when it was lowered at 0.167 s /V (open triangles).

(a)   (b)

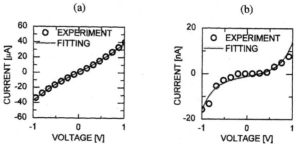

Figure 3. Experimental and fitting current-voltage carves of (a) on-state and (b) off-state of Cu nanogap junction.

Table I. Parameters estimated from the fitting curves in Fig. 3.

	AREA [nm²]	WORK FUNCTION [eV]	GAP SIZE [nm]
OFF STATE	0.49	0.54	1.02
ON STATE	180	0.66	0.6

Figure 4 shows the memory effect of Cu nanogap junction. The resistances shown in Fig. 4(a) were measured at 0.2 V after the application of bias voltages shown in Fig. 4(b). In the cycle, the resistance state immediately after application from -1 to 10 V (0.167 s /V) (wave A) is referred as the off-state, and that immediately after application from 10 to -1 V (0.167 s /V) with a current compliance of 1.0 mA (wave B) is referred to as the on-state. The current compliance was applied to prevent nanogap junctions from breaking by unexpected large current flows. A few data points show a poor on-off ratio; however, the majority of points indicate a large current on-off ratio of over 10^5. Since the junction exhibited on and off capacities after cycling over 1000 times, the Cu nanogap junction fabricated in a nanowire also exhibits memory effect with a large on-off ratio like the nanogap junctions with Au electrode.[1] In addition, Cu nanogap junctions did not obtain enough repetition (under 10 times) cycle at 2×10^{-2} Pa of vacuum level. The dependence of vacuum pressure suggests that effect of oxidation on Cu electrode like as a Si nanogap junction.[5]

(a)   (b)

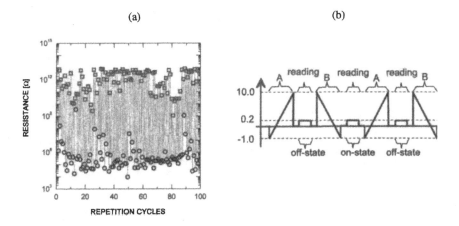

Figure 4. (a) Memory effect in Cu nanogap junction. Measurements were taken at a voltage of 0.2 V. On- and off-states were performed using the diagram of bias voltages in (b).

CONCLUSIONS

In conclusion, we investigated the electrical properties of Cu nanogap junction. The nanogap junctions with Cu electrodes show stable resistance switching under a high vacuum condition of 1.8×10^{-4} Pa. The memory effect in Cu nanogap junction shows a large on-off ratio of over 10^5. On the other hand, the effect did not obtain enough repetition cycles at 2×10^{-2} Pa of vacuum level. This suggests that the dependence on vacuum level is caused by the oxidation of electrode surface. It is implied that fresh Cu nanogap junctions show stable memory effect like as other-metal nanogap junctions.

ACKNOWLEDGMENTS

This research was supported by a grant from the New Energy and Industrial Technology Development Organization (NEDO) Innovation Research Project on Nano-electronics Materials and Structures. The authors also extend their deep appreciation to Mr. Y. Masuda, Mr. S. Furuta, and Dr. M. Ono of Funai Electronics Advance Applied Technology Research Institute, Inc., and to Dr. K. Somu of the Nanotechnology Research Institute, National Institute of Advanced Industrial Science and Technology (AIST), for their advice on a wide range of issues regarding this work. Fabrication in this work was portly supported by the Nano-Processing Facility (NPF) in AIST.

REFERENCES

1. Y Naitoh, M Horikawa, H Abe, and T Shimizu, Nanotechnology **17,** 5669 (2006).
2. S. Furuta, T. Takahashi, Y. Naitoh, M. Horikawa, T. Shimizu, and M. Ono, Jap. J. Appl. Phys. **47,** 1806 (2008).
3. Y. Masuda, T. Takahashi, S. Furuta, M. Ono, T. Shimizu, and Y. Naitoh, Appl. Surf. Sci. **256,** 1028 (2009).
4. Y. Li, A. Sinitskii, and J. M. Tour, Nat. Mater. **7,** 966 (2008).
5. Y. Naitoh, Y. Morita, M. Horikawa, H. Suga, and T. Shimizu, Appl. Phys. Express. **1,** 103001 (2008).
6. The International Technology Roadmap for Semiconductors, Semiconductor Industry Association, San Jose, CA 2005.
7. D. R. Strachanan, D. E. Smith, D. E. Johnston, T. H. Park, M. J. Therien, D. A. Bonnell, and A. T. Johnsonb, Appl. Phys. Lett. **86,** 043109 (2005).
8. G. Esen and M. S. Fuhrer, Appl. Phys. Lett. **87,** 263101 (2005).
9. H. Heersche, G. Lientschnig, K. O'Neill, H. Zant, and H. Zandbergen, Appl. Phys. Lett. **91,** 072107 (2007).
10. F. O. Hadeed and C.Durkan, Appl. Phys. Lett. **91,** 123120 (2007).
11. G. Simmons, J. Appl. Phys. **34,** 1793 (1963).

Mater. Res. Soc. Symp. Proc. Vol. 1250 © 2010 Materials Research Society
1250-G12-03

Analysis on Resistance Change Mechanism of NiO-ReRAM Using Visualization Technique of Data Storage Area with Secondary Electron Image

K. Kinoshita[1,2], T. Makino[1], T. Yoda[1], K. Dobashi[1], and S. Kishida[1,2]

[1]Department of Information and Electronics, Graduate School of Engineering, Tottori University, 4-101 Koyama-Minami, Tottori 680-8552, Japan.
[2]Tottori University Electronic Display Research Center, 522-2 Koyama-Kita, Tottori 680-0941, Japan.

ABSTRACT

Both a low and a high resistance states which were written by voltage application in a local region of NiO/Pt films by using conducting atomic force microscopy (C-AFM) were observed by using scanning electron microscope (SEM) and electron probe micro analysis (EPMA). The writing regions are distinguishable as dark areas in a secondary electron image and thus can be specified without using complicated sample fabrication process to narrow down the writing regions such as the photolithography technique. In addition, the writing regions were analyzed by using energy dispersive X-ray spectroscopy (EDS) mapping. No difference between the inside and outside of the writing regions is observed for all the mapped elements including C and Rh. Here, C and Rh are the most probable candidates for contamination which affect the secondary electron image. Therefore, our results suggested that the observed change in the contrast of the secondary electron image is related to the intrinsic change in the electronic state of the NiO film and the secondary electron yield is correlated to the physical properties of the film.

INTRODUCTION

Recent years, research and development of resistive random access memory (ReRAM) which utilizes voltage induced resistive change in transition metal oxides (TMOs) as memory media is advancing. ReRAM is a candidate for a substitution for the Flash memory, which is facing a micro-fabrication limit in the near future. Furthermore, a development of a nonvolatile and high-density universal memory with fast switching and high switching endurance is expected in the future. However, the optimization of the performance and the establishment of reliability have been prevented by the facts that switching mechanism of ReRAM has not yet been clarified. Therefore, the elucidation of switching mechanism is urgently required.

ReRAM has a simple structure of a top electrode (TE)/TMO/a bottom electrode (BE). Memory effect develops after a forming process, and it becomes possible to cause a set, which is a resistive switching from a high resistance state (HRS) to a low resistance state (LRS), and a reset, which is a resistive switching from the LRS to the HRS, alternately. Here, the forming is a phenomenon, which is similar to a soft breakdown, and a conductive path called a filament is formed after it. Resistive switching is thought to take place in the filament [1-4]. The reason why the elucidation of the resistive switching mechanism is hindered is attributed to the difficulty in applying conventional analytical methods to the resistance switching region due to the facts that the filament is covered with electrodes and that the radius of the filament is very small [5,6].

Recently, it was demonstrated that both the LRS and the HRS can be written over an arbitrary area by applying dc bias voltage, V_{bias}, directly to NiO films using conducting atomic force microscopy (C-AFM) [7,8]. The resistance writing region consists of the aggregation of filaments and can be regarded as the one filament with large radius. Therefore, this technique can provide the filament with arbitrary radius without TE, which allows application of conventional analytical methods, and might be a breakthrough for the switching mechanism elucidation.

In this paper, the LRS and the HRS with the large area of 20×20 μm^2 were written in a NiO/Pt film by voltage application using C-AFM. Taking advantage of the largeness of the target area, the writing regions were analyzed by using the scanning electron microscopy (SEM), electron micro probe (EPMA), and energy dispersive X-ray spectroscopy (EDS) for the elucidation of a physical property of the filament.

EXPERIMENTAL

A 60-nm NiO film was deposited on a Pt(100nm)/Ti(20nm)/SiO$_2$(100nm) substrate by using DC reactive magnetron sputtering at 380 °C in the mixture gas of Ar and O$_2$ gases (Ar + O$_2$ gas), and a NiO/Pt/Ti/SiO$_2$ (NiO/Pt) structure was obtained. Here, a Ni metal with the purity of 99.99% was used as a sputtering target. During the deposition, the pressure of Ar + O$_2$ gas was retained at 0.5 Pa (Ar : O$_2$ = 0.45 Pa : 0.05 Pa).

For the C-AFM measurements, a Rh-coated Si tip was grounded, and the V_{bias} was applied to the BE. HRS or LRS was written in the central 20×20 μm^2 area by scanning the tip under V_{bias}'s of -7 V or +7 V, respectively. Subsequently, the 22×22 μm^2 area containing the writing region was scanned with a sensing voltage of +1 V. The HRS was written with a negative bias, whereas LRS with a positive bias as reported in ref. [7]. The scanning frequency was 1Hz, and the data density was 256×256.

Analyses with EPMA and SEM on the regions where the HRS and the LRS were written were conducted. EDS mapping analysis on areas containing these writing regions were also performed.

RESULTS AND DISCUSSION
Resistance states written by using C-AFM

Figures 1(a) and 1(b) show current-voltage (I-V) characteristics of AFM-tip/NiO/Pt structures measured by positioning the tip at the arbitrary selected point, and applying V_{bias}'s of +7 V and -7 V, respectively. Here, the NiO film is in the initial state in both Figs. 1(a) and 1(b). It was confirmed that the HRS was written by the application of a negative V_{bias}, whereas the LRS was written by the application of a positive V_{bias}, which is consistent with ref. [7]. This effect is hard to be interpreted as an anodization of the NiO surface, which is caused by the electrolysis of water between the NiO surface and the AFM-tip [9], since reset occurs by applying a negative V_{bias} to the BE, which is the opposite bias polarity needed for the anodization to occur.

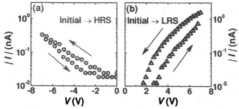

Fig. 1 I-V characteristics measured with C-AFM by application of V_{bias}'s of (a) +7 V and (b) -7 V, respectively.

Figures 2(a) and 2(b) show current images before and after performing C-AFM writing under V_{bias} of +7 V, respectively. On the other hand, Figs. 2(c) and 2(d) show current images before and after performing C-AFM writing under V_{bias} of -7 V, respectively. The bright contrast regions in the current image correspond to the LRS, and the dark contrast regions to the HRS.

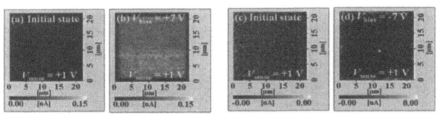

Fig. 2 Current images (a) before and (b) after C-AFM writing by scanning under V_{bias} of +7 V. Current images (c) before and (d) after C-AFM writing by scanning under V_{bias} of -7 V.

SEM and EPMA analyses

Figure 3(a) shows the locations of four 20 x 20 μm^2 regions on the NiO film to which resistance states were written by using C-AFM. Resistance states of the regions (1), (2), (3), and (4) were written by scanning the AFM-tip under V_{bias}'s of +7 V, -7 V, +5 V, and -5V, respectively. Regions (1) and (3) were written to the LRS, whereas regions (2) and (4) to the HRS. That is, the resistance decreased by application of positive V_{bias}, whereas increased by negative bias V_{bias}, as described above. The scratch in the figure was introduced by a diamond-point pencil to support finding out the same place containing these regions by EPMA and SEM.

Figures 3(b), 3(c), and 3(d) represent a secondary electron (SE), a composition (COMP), and a topographic (TOPO) images of the NiO film containing all the regions (1)-(4), respectively. Here, COMP and TOPO images are obtained by a primary electron emission, which is sensitive to both the composition and surface morphology of the sample, respectively. Dark regions in Fig. 3(b), where a secondary electron yield is low compared with the other region, correspond to the regions (1)-(4), whereas no difference in the contrast from the surrounding region is observed in the corresponding region in Figs. 3(c) and 3(d). In addition, the SE image becomes darker with increasing the amplitude of writing voltage.

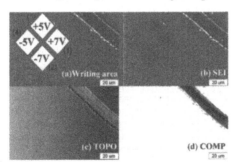

Fig. 3 (a) Schematic that shows location of regions (1)-(4) written by scanning under V_{bias}'s of +7, -7, +5, and -5 V using C-AFM, respectively. (b) SE, (c) TOPO, and (d) COMP images of the NiO film in which regions (1)-(4) are included.

229

Figures 4(a) and 4(b) show SE images before and after the C-AFM writing (not the same position). Tetrahedral grain structures are observed on the surface, showing the (111) orientation of the NiO film. No remarkable change in both sizes and shapes of NiO grains were observed. No damage caused by the writing current was also confirmed.

Figure 5 shows applied bias and scanning frequency dependences of a SE image for writing regions. In Fig. 5, regions (1)-(3) were written to the LRS and regions (4)-(6) were written to the HRS under the writing conditions shown in table. 1. The higher the applied bias becomes and/or the lower the scanning frequency becomes, the darker the SE image of the writing regions becomes. Therefore, a darker SE image is obtained by applying V_{bias} with larger amplitude for a longer duration to the unit area.

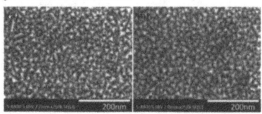

Fig. 4 SE images before and after C-AFM writing under V_{bias} of +7V with a scanning frequency of 0.2 Hz.

Table. 1 Conditions under which regions (1)-(6) in Figs. 5 and 6(a) were written to the HRS or the LRS by using C-AFM.

Writing regions	Writing-V V [V]	Frequency ν [Hz]
(1)	+9	1
(2)	+9	0.2
(3)	+5	1
(4)	-9	1
(5)	-9	0.2
(6)	-5	1

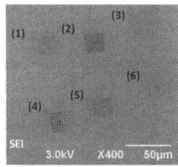

Fig. 5 SE image after C-AFM writing for regions (1)-(6) under the conditions shown in table. 1.

EDS mapping analysis

As possible main factors which cause the change in the contrast of a SE image by the application of V_{bias} using C-AFM, the following three factors can be pointed out: (i) The change in the surface morphology has been caused. (ii) Removal/adhesion of contaminations from/to the surface of the NiO film has been caused. (iii) An intrinsic change in electronic state of NiO has been caused. Since no remarkable change in the surface morphology of the sample by the application of C-AFM writing was observed, which denied the factor (i), the possibility of the factor (ii) will be discussed below.

Elements for which EDS mapping analysis were conducted were selected as follows. Hilleret *el al.* reported that secondary electron yield of an air exposed ZnO film decreased after

230

annealing the film [10]. This is due to a removal of airborne carbonaceous contamination layer formed on the surface of the sample by the annealing process. Analogously, also in the present study, the removal of carbonaceous contamination might be caused by Joule heating generated during C-AFM writing. Therefore, C was selected as a mapping element. In addition, Rh with which the AFM-tip is coated might adhere to the scanned area. By adding the constituent elements of the sample to C and Rh, EDS mapping analyses were performed on C, Rh, Ni, O, and Pt.

Figure 6(a), which is the same figure as Fig. 4, shows the SE image after the C-AFM writing. The square regions surrounded by dotted lines correspond to the writing regions which were written under the conditions shown in Table. 1. Figures 6(b)-6(f) show results of EDS mapping measurements for Rh, C, O, Ni, and Pt, respectively. Here, the square regions surrounded by dotted lines in Fig. 6(b)-6(f) correspond to the writing regions in Fig. 6(a), respectively. The acceleration voltage was 3.0 kV. Energies of C K_α, O K_α, Ni L_α, Pt M_α, and Rh L_α lines are 0.277, 0.525, 0.851, 1.739, 2.048, and 2.696 eV, respectively.

No difference of the intensity distribution in the EDS mappings between inside and outside of the writing regions was observed for all the mapped elements, suggesting the exclusion of the factor (ii). Therefore, as a reason why the SE image becomes dark by the voltage application, it was suggested that the change in the electric conductivity of NiO affected

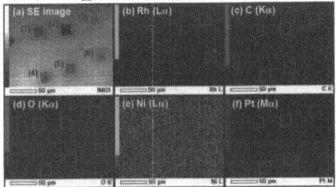

Fig. 6 (a) Secondary electron image after C-AFM writing under the conditions shown in table. 1, which is the same as Fig. 5. EDS mapping results for (b) Rh, (c) C, (d) O, (e) Ni, and (f) Pt.

the secondary electron yield.

It has been widely received that the resistance change effect observed in transition metal oxides is caused by the migration of oxygen ions [2,7,11], which introduces or repairs oxygen vacancies depending on the applied bias condition. It has also been received that the LRS is formed by ranging vacancies through the film, which is the so-called filament, whereas HRS is formed by repairing the vacancies only near the electrode interface [2,12,13]. This means that conductive filaments other than the neighborhood of the electrode interface remain without being repaired even in the HRS. Therefore, this is consistent with the fact that the writing region becomes dark whether it is in the HRS or the LRS in the SE image if assuming that a conductive region becomes dark in the SE image. Shima et al. performed Kelvin probe force microscopy and micro X-ray photoelectron spectroscopy measurements on the LRS of a CoO film which was written by the voltage application using C-AFM [6]. They reported that Fermi energy in CoO is pushed up because of reduction of CoO which causes defect-related energy levels in the band gap. Since it was reported that low secondary electron yield was related to a low work function [14], our result also suggests that the C-AFM writing decreases a work function, which is consistent with ref. [6]. However, in ref. [6], since a writing region is optically unidentifiable, the region excluding the region for which C-AFM writing will be performed is etched by using the photo lithography to narrow down the writing region from the whole region to analyze the writing region by XPS. Therefore, it is worth noting that we obtained the result similar to that reported in ref. [6] without the influence of the photo lithography on the analyzed area of the sample.

CONCLUSIONS

SEM and EPMA analyses on the C-AFM writing region revealed that the writing regions are distinguishable as dark areas in SE images. This enables to specify the writing region without using complicated sample fabrication process to narrow down the writing regions such as the photolithography technique. In addition, EDS mapping analysis on the writing region suggested that the observed change in the contrast of the SE image is related to the intrinsic change in the electronic state of the NiO film, i.e., that the secondary electron yield is correlated to the physical properties of the film.

ACKNOWLEDGMENTS

The authors thank H. Kasada for his experimental support.

REFERENCES

1. I. G. Baek, M. S. Lee, S. Seo, M. J. Lee, D. H. Seo, D.-S. Suh, J. C. Park, S. O. Park, H. S. Kim, I. K. Yoo, U-In Chung and J. T. Moon, Tech. Digest IEDM 2004, p. 587.
2. K. M. Kim, B. J. Choi, and C. S. Hwang, Appl. Phys. Letts. **90**, 242906 (2007).
3. K. Kinoshita, T. Tamura, M. Aoki, Y. Sugiyama, and H. Tanaka, Jpn. J. Appl. Phys. **45**, L991 (2006).
4. J. F. Gibbons and W. E. Beadle, Solid-State Electron. **7**, 785 (1964).
5. M.-J. Lee, S. Han, S. H. Jeon, B. H. Park, S. Kang, S-E. Ahn, K. H. Kim, C. B. Lee, C. J. Kim, I.-K. Yoo, D. H. Seo, X.-S. Li, J.-B. Park, J.-H. Lee, and Y. Park, Nano Lett. 9, 1476 (2009).
6. H. Shima, F. Takano, H. Muramatsu, M. Yamazaki, H. Akinaga, and A Kogure, phys. stat. sol. (RRL) **2**, 99 (2008).
7. C. Yoshida K. Kinoshita, T. Yamasaki, and Y. Sugiyama, Appl. Phys. Lett. **93**, 042106 (2008).
8. K. Kinoshita, T. Okutani, H. Tanaka, T. Hinoki, K. Yazawa, K. Ohmi, and S. Kishida, Appl. Phys. Lett. **96**, 143506 (2010).
9. J. Hsu, H. Lai, H. Lin, C. Chuang, and J. Huang, J. Vac. Sci. Technol. B **21**, 2599 (2003).
10. N. Hilleret, C. scheuerlein, M. taborelli, Appl. Phys. A **76**, 1085 (2003).
11. Y. B. Nian, J. Strozier, N. J. Wu, X. Chen, and A. Ignatiev, Phys. Rev. Lett. **98**, 146403 (2007).
12. K. Kinoshita, T. Tamura, M. Aoki, Y. Sugiyama, and H. Tanaka, Appl. Phys. Lett. **89** (2006) 103509.
13. I. H. Inoue, S. Yasuda, H. Akinaga, and H. Takagi, Phys. Rev. B **77** (2008) 035105.
14. M. Kudo, Y. Sakai and T. Ichinokawa, Appl. Phys. Lett. **76**, 3475 (2000).

Mater. Res. Soc. Symp. Proc. Vol. 1250 © 2010 Materials Research Society   1250-G12-12

Electrical Properties of TaO$_x$ Thin Films for ReRAM Prepared by Reactive RF Magnetron Sputtering Method

Natsuki Fukuda, Hidenao Kurihara, Kazumasa Horita, Yoshiaki Yoshida, Yutaka Kokaze, Yutaka Nishioka, and Koukou Suu
Institute of Semiconductor and Electronics Technologies, ULVAC, Inc.
1220-1 Suyama, Susono, Shizuoka, 410-1231, Japan
natsuki_fukuda@ulvac.com

ABSTRACT

TaO$_x$ thin films were fabricated by O$_2$ reactive Radio Frequency (RF) magnetron sputtering on 8inch substrate using a Ta metal target at room temperature for mass-production of ReRAM. The TaO$_x$ thin films had good uniformity ($\pm$1%), excellent stability($\pm$1%) at the sputtering process. Ta/TaO$_x$(10nm)/Pt-ReRAM cell was confirmed the bipolar switching and excellent endurance property to 7×10^8 cycles with large on/off resistance ratio above 1,000 at high speed operation (50nsec, $\pm$3V).

INTRODUCTION

Recently NAND Flash memory will be scaling limit under 25nm technical node [1], so the next generation non-volatile memory is actively researched. Resistive Random Access Memories (ReRAM) has advantages of scaling, high speed and low power operation in comparison with the other non-volatile memories.

The first report of resistance switching was due to T. W. Hickmott et al. in 1962, the materials are SiO$_2$, Al$_2$O$_3$, Ta$_2$O$_5$, ZrO$_2$ and TiO$_2$, it's the preparation methods were evaporation and anodization [2]. 64kbit-array for memory application were reported by W. W. Zhuang in 2002, (Pr,Ca)MnO$_3$ (PCMO) deposited on Si substrate by Metal Organic Deposition (MOD) and Pulse Laser Deposition (PLD) methods [3].

The materials for ReRAM are classified into the perovskite oxides and the binary oxides. The perovskite oxides such as PCMO are fabricated at high temperature over 700℃ [3], [4], [5]. On the other hand, the binary oxides such as NiO$_x$ [6], TiO$_x$ [7] and HfO$_x$ [8] are fabricated at low temperature under 400 ℃. Moreover, binary oxides are highly compatible with a conventional CMOS process. So binary oxides have many advantage for mass-production.

The sputtering method has many advantages of throughput, thickness uniformity for large substrate, stability of deposition, low temperature and low cost for ReRAM mass-production compare with the other preparation method.

In this study, we prepared TaO$_x$ thin films by O$_2$ reactive RF magnetron sputtering and evaluate the uniformity, stability and resistive switching performance.

EXPERIMENT

Cluster type sputtering tool (ULVAC, CERAUS ZX-1000, [9], [10]) used in this experiment is illustrated in figure 1. TaO_X thin films were prepared by O_2 reactive RF magnetron sputtering on 8inch-Pt (100nm)/Si substrate using a Ta metal target at room temperature. The properties of TaO_X film, thickness, refractive index, particle, and crystal phase were evaluated by ellipsometer (ESM-1AT, ULVAC), particle counter (SP1, KLA tencore) and X-ray diffraction meter (MRD, PANalytical). After deposited TaO_X thin film, 100nm-thick Ta or Pt top electrode were formed by DC magnetron sputtering method through a shadow mask with dots of 50 μ m in diameter. Switching and endurance properties of (Pt or Ta)/TaO_X(10nm)/Pt cell were measured by semiconductor device analyzer (B1500A, Agilent) and pulse generator (81110A, Agilent).

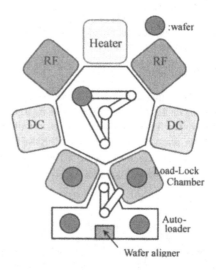

Figure 1. Sputtering system for mass-production (CERAUS ZX-1000, ULVAC, Inc.)

RESULTS AND DISCUSSION

1. Uniformity and Stability

Figure 2 shows uniformity of the thickness and the refractive index of TaO_X film by ellipsometer (λ =633nm) in 8inch wafer. Arithmetic average of the thickness and the refractive index were 60.1nm and 2.08 respectively. Uniformities of the thickness and the refractive index were $\pm$0.9% and $\pm$1.1% respectively.

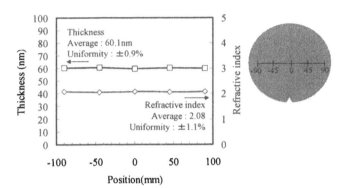

Figure 2. Uniformity of thickness and refractive index of TaO_X films by ellipsometer in 8inch wafer.

Stability of the deposition rate and the particle were indicated in figure 3 and figure 4 respectively. TaO_X films were prepared without the maintenance of as many as about 90 wafers in the thickness of 60 nm. Reproducibility of the rate by arithmetic average was 16.2nm/min, and uniformity was ±1.0%. Also stability of the particle over 0.2 μ m size was under 40 pieces.

Excellent uniformity and process stabilities of TaO_X films were confirmed by the sputtering reproducibility test.

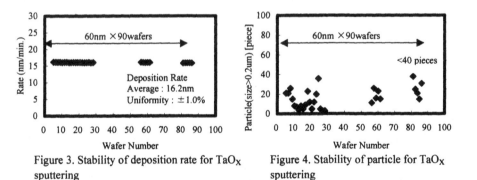

Figure 3. Stability of deposition rate for TaO_X sputtering

Figure 4. Stability of particle for TaO_X sputtering

2. Crystallinity

Figure 5 shows the XRD pattern of 60nm TaO_X films on Pt/Si substrate. TaO_X crystal structure was not observed by XRD.

235

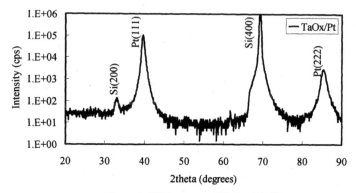

Figure 5. XRD pattern of TaOx thin films

3. Switching properties

Figure 6 shows the Current - Voltage characteristics of (a) Pt/TaO$_X$(10nm)/Pt cell and (b) Ta/TaO$_X$(10nm)/Pt cell. The initial state was high resistance to either cell [A and A'], and when the voltage of 4.5V or more was applied, the soft breakdown was caused with a current compliance program of B1500A [B and B']. Afterwards, when –2.5V were applied to (b) Ta/TaO$_X$(10nm)/Pt cell, it changes from the low resistance state (off state) to the high resistance state (on state) [D'→E']. However, even if minus voltage was applied, the change in resistance was not observed on (a) Pt/TaO$_X$(10nm)/Pt cell. It suggests that resistance switching occurs by the interface between the top electrode and the oxide.

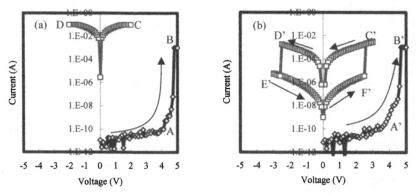

Figure 6. Current – Voltage characteristics of (a) Pt/TaO$_X$ (10nm)/Pt cell and (b) Ta/TaO$_X$ (10nm)/Pt cell

Figure 7 shows the uniformity of the switching property of Ta/TaO$_X$(10nm)/Pt cell in 8inch wafer. The switching properties of all position (center, middle, edge) had an approximately ±3V operation voltage and similar current value.

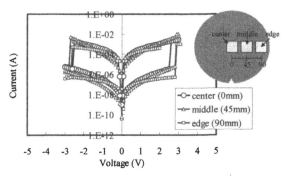

Figure 7. Switching properties of Ta/TaO$_X$(10nm)/Pt cell in 8inch wafer.

Figure 8 shows the endurance property of Ta/TaO$_X$(10nm)/Pt cell. The set (off→on) and reset (on → off) pulse voltages were 3.0V 50nsec and -3.0 V 50nsec respectively. Ta/TaO$_X$(10nm)/Pt cell had an excellent endurance property to 7×10^8 cycles with large on/off resistance ratio above 1,000 at high-speed operation.

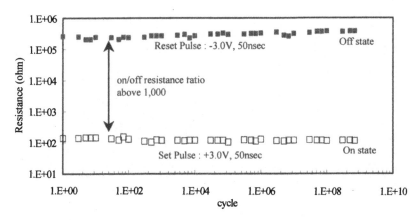

Figure 8. Endurance property of Ta/TaO$_X$(10nm)/Pt cell

CONCLUSIONS

TaO_x thin films were prepared by O_2 reactive RF magnetron sputtering method using a Ta metal target at R.T. for mass-production of ReRAM. The TaO_x films had a good uniformity and stability of the thickness and low particle level. $Ta/TaO_x(10nm)/Pt$ cell was confirmed the bipolar switching, and good uniformity of it. At high speed operation (50nsec, $\pm 3V$), the ReRAM cell was indicate the excellent endurance property to 7×10^8 cycles with large on/off resistance ratio above 1,000.

ACKNOWLEDGMENTS

This work was supported by NEDO (New Energy and Industrial Technology Development Organization in Japan).

REFERENCES

[1] International Technology Roadmap for Semiconductors 2009
[2] T. W. HICKMOTT: J. Appl. Phys. 33, 2669 (1962)
[3] W. W. Zhuang, W. Pan, B. D. Ulrich, J. J. Lee, L. Stecker, A. Burmaster, D. R. Evans, S. T. Hsu, M. Tajiri, A. Shimaoka, K. Inoue, T. Naka, N. Awaya. Sakiyama, Y. Wang, S. Q. Liu, N. J. Wu and A. Ignatiev: Tech. Dig. Int. Electron Devices Meet., San Francisco, 2002, p193
[4] A. Beck, J. G. Bednorz, Ch. Gerber, C. Rossel and D. Widmer: Appl. Phys. Lett. 77, 139 (2000)
[5] A. Odagawa, H. Sato, I. H. Inoue, H. Akoh, M. Kawasaki, and Y. Tokura, T. Kanno and H. Adachi: Phys. Rev., B 70, 224403 (2004)
[6] I. G. Baek, M. S. Lee, S. Seo, M. J. Lee, D.H. Seo, D.-S. Suh, J. C. Park, S. O. Park, H. S. Kim, I. K. Yoo, U-In Dhung and J. T. Moon: Tech. Dig. Int. Electron Devices Meet., San Francisco, 2004, 23, 6, p587
[7] B. J. Choi, D. S. Jeong, S. K. Kim, C. Rohde, S. Choi, J. H. Oh, H. J. Kim, C. S. Hwang, K. Szot, R. Waser, B. Reichenberg, S. Tiedke: J. Appl. Phys. 98, pp033715 (2005)
[8] Heng-Yuan Lee, Pang-Shiu Chen, Ching-Chiun Wang, Siddheswar Maikap, Pei-Jer Tzeng, Cha-Hsin Lin, Lurng-Shehng Lee, and Ming-Jinn Tsai: Extended Abstracts of the 2006 International Conference on Solid State Devices and Materials, Yokohama, 2006, pp.288-289
[9] Koukou Suu, Yusuke Miyaguchi, T.Masuda, Y.Nishioka and Fan Chu: TECHNICAL REPORT OF IEICE. ED2000-69, SDM2000-69 (200-06), p49-56
[10] TAKEHITO JIMBO, YUUSUKE MIYAGUCHI, SHIN KIKUCHI, ISAO KIMURA, MICHIO TANIMURA, KOUKOU SUU and MICHIO ISHIKAWA: Integrated Ferroelectrics, 2002, Vol. 40, pp.105-112

Mater. Res. Soc. Symp. Proc. Vol. 1250 © 2010 Materials Research Society　　　　1250-G12-16

Characteristics of ZnO Thin Film for the Resistive Random Access Memory

Jung Won Seo, Seung Jae Baik, Sang Jung Kang and Koeng Su Lim
Department of Electrical Engineering, Korea Advanced Institute of Science and Technology
(KAIST), 335 Gwahak-ro, Yuseong-gu, Daejeon 305-701, Republic of Korea

ABSTRACT

We report resistive switching characteristics in Pt/ZnO/Pt devices where ZnO thin film is fabricated at various oxygen conditions. With the increase of oxygen contents in ZnO thin film, the forming voltage is gradually increased while reset and set voltages are almost unchanged. We also investigated the effect of top electrodes on resistive switching of top electrode/ZnO/Pt device. For a Pt/ZnO/Pt device, it exhibits the excellent resistive switching behavior due to high electrical field of the well-defined Schottky barrier. For Al/ZnO/Pt device, little resistive switching phenomena were occurred due to leakage current of the weak Schottky (or Ohmic) contact.

INTRODUCTION

Currently resistive random access memory (RRAM) is an emerging class of device for next-generation nonvolatile memory device due to its simple structure, good scalability and compatibility with complementary metal oxide semiconductor technology. The resistive switching behavior has been widely observed in ferromagnetic oxide materials ($Pr_{1-x}Ca_xMnO_3$) [1], doped perovskite oxide materials ($SrZrO_3$) [2], and organic materials (poly-N-vinylcarbazole) [3], as well as in simple binary transition metal oxide materials [4].

Especially, various electrical properties have been reported in metal/ZnO/metal device and explained with resistive switching mechanism such as conducting filament [5], space charge limited conduction (SCLC) [6] and Poole-Frenkel emission model [7]. However, comprehensive understanding for ZnO nature on resistive switching behavior is lacking to fulfill the requirement for the commercialization as a type of next-generation memory application. Therefore, the fundamental analyses such as oxygen contents of ZnO thin film and effects of electrodes on the switching properties must be investigated systematically as an important parameter in order to realize of ZnO thin film RRAM device.

In this work, we fabricated ZnO thin film with various oxygen contents (0~50 %) and introduce the resistive switching behavior of Pt/ZnO/Pt device. Also we demonstrated the effect of electrode (Al, Cr, Au and Pt) on resistive switching properties of top metal/ZnO/Pt device.

EXPERIMENTAL DETAIL

The fabricated Pt/ZnO/Pt device is shown in the inset of Fig. 1(c). About 50 nm ZnO thin films were deposited on Pt/Ti/SiO$_2$/Si substrates by rf magnetron sputtering method at room temperature. During deposition process, the total flow rate of Ar and O_2 is fixed to 20 sccm and the O_2 contents (Co_2) is varied from 0 to 50 % [Co_2 (%) = $c(O_2) / c(Ar+O_2)$] . The process pressure and rf power were kept at 5 mTorr and 200 W, respectively. As a top electrode, a 100 nm thick Al, Cr, Au and Pt were deposited by dc sputtering with a metal shadow mask and were ranged from 50 to 500 μm in diameter. The current (*I*)-voltage (*V*) characteristics are measured with an HP 4156A semiconductor parameter analyzer and a 4150B pulse generator. During the *I*-*V* measurement, the top electrode is biased positively and the bottom electrode is always grounded. For the fundamental analyses of the ZnO thin film, we performed x-ray diffraction (XRD), x-ray photoelectron spectroscopy (XPS), auger electron spectroscopy (AES), atomic force microscopy (AFM) and Rutherford backscattering spectroscopy (RBS) (in this paper, AFM and RBS results were shown only).

DISCUSSION

Figure 1(a) shows the RBS and AFM (inset) results of ZnO thin film. We observed that the surface of ZnO thin film is very uniform (rms: < 1 nm) regardless of the oxygen contents (not shown all sample). With the RBS spectrum, we analyzed the composition of ZnO thin film qualitatively, resulting that the ratio of O on Zn(1) is varied with 0.68 (at 0 % O_2 content), 0.73 (at 10 % O_2 content), 0.74 (at 20 % O_2 content), and 0.75 (at 50 % O_2 content).

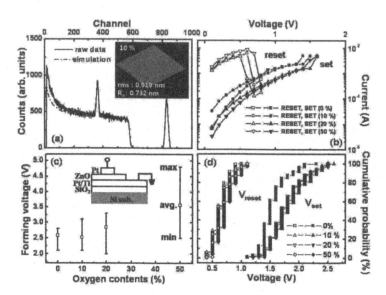

Figure 1. (a) AFM and RBS spectrum of ZnO thin film. Scanned area is 5 x 5 μm². (b) Resistive switching of Pt/ZnO/Pt device with various oxygen contents. Distribution of (c) forming voltage and (d) set and reset voltage (d) with oxygen contents. The device sizes are all 200 μm.

The resistive switching characteristics of Pt/ZnO/Pt devices are investigated, as shown in Fig. 1(b). After forming process, the Pt/ZnO/Pt device switched to a low resistance state (LRS). The device then reached a high resistance state (HRS) when the voltage swept from zero to positive value (RESET process). By re-sweeping the positive voltage, the device switched back to the LRS (SET process) and nonvolatile switching was accomplished. We observed that the switching parameters (V_{set}, V_{reset}, I_{set} and I_{reset}) of the devices are very uniform, without regard to the oxygen contents of ZnO thin film. The ON (LRS)/OFF (HRS) ratio is about 2 orders magnitude. The dependence of the forming voltage of the device on the oxygen contents of ZnO thin film also inspected. As can be seen in Fig. 1(c), the forming voltage is increased with the oxygen contents. In general, a forming process with current compliance (here, 5 mA) to prevent fatal destruction of the device is necessary to activate the nonvolatile switching properties of a RRAM device [8]. It is well-known that the conductive filament is mainly composed of oxygen vacancies in the ZnO thin film and has an important role in resistive switching. During the forming process, oxygen ions in the ZnO thin film were pulled into the top electrode by the electric field (forming voltage), forming the oxygen vacancies, the empty location of oxygen ions, then conductive filament is evolved by aligning these oxygen vacancies. Therefore, the forming voltage to make the conductive filament is increased with the increase of the oxygen contents because the concentration of oxygen vacancies of ZnO with high oxygen content is lower than that with low oxygen content. Fig. 1(d) exhibits the cumulative distribution of reset and set voltage in the Pt/ZnO/Pt device. For clear comparison, about 130 ~ 209 times measurements (for 9 devices, including consecutive switching) were tested. The results show that the extent of reset voltage (V_{reset}, 0.4 ~ 1 V) is more uniform than that of set voltage (V_{set}, 1.1~ 2.3 V). It is reported that SET process should be more random than the RESET process, since the formation process is determined by the competition among different filamentary paths. Besides, it is known that set process is similar with forming process in that it also formed the conductive filament. But, V_{set} is not seriously affected with the oxygen contents unlike the forming voltage. It is attributed that the formation of the conductive filament by the set process is only occurred at ruptured conductive filament. During the reset process, the conductive filament around the anode is ruptured by large current flow, presumably joule heating effect, so most of the conductive filaments are remained. Set process is to make the conductive filament only at ruptured small part of conductive filament by the reset process whereas forming process is to make the whole conductive filament through ZnO thin film. Therefore, distribution of V_{set} is nearly independent of concentration of oxygen contents in ZnO thin film.

Dependence of temperature, device size and current compliance

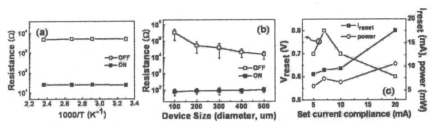

Figure 2. (a) Temperature dependence and (b) area dependence of LRS and HRS. (c) Set current compliance dependence of reset voltage and reset current. The device size in (a) and (c) are 300 μm.

We investigated the temperature dependence of the resistance in the LRS and HRS of the Pt/ZnO/Pt device. As shown in Fig. 2(a), the resistances of LRS and HRS are unchanged up to 150 °C without any degradation, indicating that the device exhibits very excellent thermal stability. We also examined the cell area dependence of the resistance in LRS and HRS, shown in Fig. 2(b). The resistance of the HRS is increased with decrease of cell area while that of LRS is almost same regardless of cell area, which means that the current in the LRS flowed through the localized path, conductive filament, and current flow in HRS is homogeneous for whole cell area. The relationship between the compliance set current and V_{reset}, I_{reset} is shown in Fig. 2(c). I_{reset} has an almost linear relation to the set current compliance, although V_{reset} is independent of the magnitude of the compliance current. It means that I_{reset} is a dominant factor for the reset process, indicating that it is governed by Joule heating effect, presumably I^2R [9]. The increase of I_{reset} indicates that LRS resistance of ZnO thin film decreases. If a higher current level is applied to the ZnO thin film, more oxygen vacancies were generated and the conductive filaments with higher density were formed. As a result, a higher I_{reset} is required to fracture these conductive filaments, resulting in the HRS of the device.

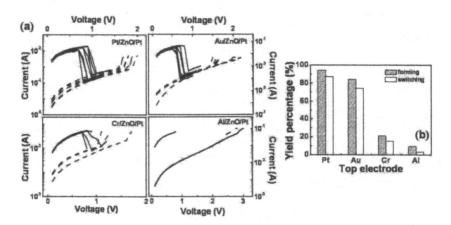

Figure 3. (a) Resistive switching properties of top metal (Pt, Au, Cr and Al)/ZnO/Pt device. The device sizes are all 200 μm. (b) Device yield with different top electrodes. The tested devices are about 50 ~ 60 (depending top electrodes).

In order to confirm the effect of electrode on switching behavior, we investigated resistive switching for metal/ZnO/Pt device with different top electrode. As can be seen in Fig. 3, the devices with Pt and Au as top electrodes exhibit the good resistive switching characteristics whereas those with Cr and Al show the poor resistive switching behavior. This result may be attributed from the difference of metal workfunction. When the metal and semiconductor make contact, the contact type is simply determined with difference between the metal workfunction and the Fermi level of semiconductor. It is well-known that ZnO is typically oxygen deficient n-type semiconductor material and has about 3.2 eV bandgap. The workfunctions of top electrodes used in this experiment are 5.12-5.93 eV (Pt), 5.1-5.47 eV (Au), 4.5 eV (Cr), and 4.06-4.26 eV (Al), respectively [10]. So the high workfunction metals (Pt, Au) easily form strong Schottky barrier with ZnO thin film while low workfunction metals (Cr, Al) may form weak-Schottky or nearly Ohmic contact. Due to the strong Schottky barrier, high electric field at contact could help the movement of oxygen ions (i.e. oxygen vacancies) during switching process and the stable resistive switching is achieved. However, low Schottky barrier or Ohmic contact could not make the high electric field, so leakage current could flow through ZnO thin film, resulting the unstable or irreversible resistive switching. These could be supported by the device yield, shown in Fig. 3(b).

CONCLUSIONS

In this paper, resistive switching behavior of Pt/ZnO/Pt device with various oxygen contents is investigated. The forming voltage is directly influenced with oxygen content whereas V_{set} and V_{reset} have little bit of effect. We also examined dependence of temperature, cell area and set current compliance on resistive switching. Through the analysis of metal/ZnO/Pt device with various top electrodes, we observed that high workfunction metal could enhance the resistive switching characteristics due to the high electric field of strong Schottky barrier. We believe that these results are expected to extend the understanding on resistive switching characteristics of ZnO RRAM.

ACKNOWLEDGMENTS

The authors are grateful to the participating lab members for their technical help. This work was supported by the Korea Science and Engineering Foundation (KOSEF) grant funded by the Korea government (MEST) (No. 2008-0062241)

REFERENCES

1. S. Q. Liu, N. J. Wu, and A. Ignatiev, *Appl. Phys. Lett.* **76**, 2749 (2000).
2. Y. Watanabe, J. G. Bednorz, A. Bietsch, Ch. Gerber, D. Widmer, A. Beck, and S. J. Wind, *Appl. Phys. Lett.* **78**, 3738 (2001).
3. Y. S. Lai, C. H. Tu, D. L. Kwong, and J. S. Chen, *Appl. Phys. Lett.* **87**, 122101 (2005).
4. M. Villafuerte, S. P. Heluani, G. Juarez, G. Simonelli, G. Braunstein, and S. Duhalde, *Appl. Phys. Lett.* **90**, 052105 (2007).
5. D. S. Jeong, H. Schroeder, and R. Waser, *Appl. Phys. Lett.* **89**, 082909 (2006).
6. J. W. Seo, J-W. Park, K. S. Lim, S. J. Kang, Y. H. Hong, J-H. Yang, L. Fang, G. Y. Sung, and H-K. Kim, *Appl. Phys. Lett.* **95**, 133508 (2009).
7. W-Y. Chang, Y-C. Lai, T-B. Wu, S-F. Wang, F. Chen, and M-J. Tsai, *Appl. Phys. Lett.* **92**, 022110 (2008).
8. J. W. Seo, J-W. Park, K. S. Lim, J-H, Yang, and S. J. Kang, *Appl. Phys. Lett.* **93**, 223505 (2008)
9. J-W. Park, J-W. Park, D-Y. Kim, and J-K. Lee, *J. Vac. Sci. Technol.* **23**, 1309 (2005).
10. D. R. Lide, *CRC handbook on Chemistry and Physics* (CRC press, Florida, 2008)

Mater. Res. Soc. Symp. Proc. Vol. 1250 © 2010 Materials Research Society   1250-G12-17

High-temperature process endurance of oxide/electrode stacking structure for resistance random access memory

Hisashi SHIMA1, Takashi NAKANO2, Hiroyuki AKINAGA1
1 Nanodevice Innovation Research Center (NIRC),
National Institute of Advanced Industrial Science and Technology (AIST),
Tsukuba Central 2, 1-1-1 Umezono, Tsukuba, Ibaraki 305-8568, Japan
2 Advanced Technology Research Laboratories, Corporate Research and Development Group,
SHARP Corporation, 1 Asahi, Daimon-cho, Fukuyama, Hiroshima 721-8522, Japan

ABSTRACT

The systematic investigation on the thermal stability of the CoO layer was carried out for various electrode materials. When Pt with higher oxygen potential (Gibbs free energy change of the oxidation reaction) compared to Co is used as electrodes, the resistance of the Pt/CoO/Pt devise was severely decreased by the post deposition annealing (PDA) process and the resistance switching into the high resistance state was observed in the first voltage sweep. This indicants that the reducing Ar ambient induces the quite local reduction of CoO. The reduction of the CoO layer is also expected even with the Co electrode, which is reasonably attributed to the oxygen concentration gradient at the Co/CoO interface in the Co/CoO/Pt device. With the Ti electrode having a much lower oxygen potential than Co, the reduction of CoO by Ti is also indicated electrically in the Pt/CoO/Ti device. On the other hands, W electrodes which is thought to have the solid-solution oxygen can stabilize the CoO layer during PDA although W is more affinitive with oxygen compared with Co. It can be pointed out the oxygen delivery at the electrode/oxide layer interface is a critical factor in designing the thermally stable stacking structure for resistance random access memory.

INTRODUCTION

Resistance random access memory (RRAM) having a bottom electrode (BE)/oxide/top electrode (TE) stacking structure attracts significant attention because of the potential to be the CMOS (complimentary metal-oxide semiconductor) compatible, ultrahigh density and ultrahigh speed non-volatile memory [1-6]. In the previous reports on the electrode material dependence of the switching characteristics, the chemical reactions between the oxide and electrode layers yielding an interfacial layer were focused on and the correlation between the switching property and the interfacial layer was discussed [4-6]. Since the interface in RRAM is a solid-solid junction, the oxidation and reduction reactions are accelerated when the junction is exposed to the elevated temperature. In view of the practical application, the switching properties should have a sufficient thermal stability because the conventional CMOS process requires RRAM to be subjected to high temperature. Here we show the degradation and the improvement of the thermal stability of RRAM depending on the electrode material. We have examined the influence of the post deposition annealing (PDA) in the reducing Ar atmosphere on the initial states of CoO RRAMs with the top electrode (TE) material of Pt, Ti, and W prepared by magnetron sputtering. It was revealed that the reduction reaction proceeds in CoO RRAM even with precious metal Pt TE. With Ti TE, the reduction of CoO by Ti is thought to be the reasonable origin for the strongly decreases of the initial resistance of the device. In contrast, CoO RRAM with W TE withstands PDA despite the stronger oxygen affinity of W compared with Co. The present results manifest that the chemical reaction unexplainable in terms of the Gibbs free energy of the oxidation and reduction processes strongly influences the stability in the RRAM devices.

EXPERIMENT

The CoO layer with the thickness of 10 nm was grown on the bottom electrode (BE) thin film by means of magnetron sputtering. The substrate is the thermally oxidized Si and BE materials beneath the CoO layer were Pt and Co. Following the CoO layer deposition, the top electrode (TE) patterns having the size of 20×20 μm^2 and 30×30 μm^2 were prepared on the BE/CoO blanket by the conventional photolithography and lift-off process. Here, the TE materials were Pt, Ti, and W. The thickness of Pt was 100 nm, while those of Ti and W were 50 nm. When TE = Ti and W, the Pt capping layer were deposited on those layers. The obtained device structures in the present paper are Pt/CoO/Pt, Pt/CoO/W/Pt, Pt/CoO/Ti/Pt, Co/CoO/Pt, and Co/CoO/W/Pt. Henceforth, for simplicity, those devices are abbreviated as PP, PW, PT, CP, and CW. In order to evaluate the resistance to the high temperature process, the post-deposition annealing (PDA) was carried out. The PDA atmosphere, temperature, and time were, respectively, purified Ar, 400 °C, and 10 min. The electrical properties of the devices were measured using Keithley 4200 semiconductor parameter analyzers, contacting W probes on BE and TE. The external voltage was driven on TE with BE grounded. The surface smoothness of the CoO layer on BE were evaluated by atomic force microscopy (AFM).

DISCUSSION

Figures 1(a) and 1(b) are the AFM images of the CoO layer on Pt BE and Co BE, respectively. The films consist of fine grains with the grain size of about 20 nm for Pt/CoO and 30 nm for Co/CoO. The average surface roughness evaluated for the scanning area of 1×1 μm^2 for Pt/CoO and Co/CoO are, respectively, around 0.6 and 0.7 nm. Almost comparable CoO films are deposited on those BEs.

Shown in Figs. 2(a) – 2(e) are examples of the resistance switching properties observed in PP, PW, PT, CP, and CW in the as prepared state, respectively. The device size used here is 20×20 μm^2. In all of those devices, three characteristic switching processes, forming, reset, and set are clearly observed. Here, the forming process is the resistance switching from the pristine state to the low resistance state (LRS). The reset process is the resistance switching from LRS to the high resistance state (HRS), while the set process corresponds to the resistance switching from HRS to LRS. In the case of PP and CP, the unipolar resistance switching was observed. On the other hands, the other devices, PT, PW, and CW showed a stable bipolar resistance switching. The current regulation during the forming and set process was carried out by using the current compliance program in the semiconductor parameter analyzer or by using the load resistor in series with RRAM. It has been reported in the previous several papers that the transient current during the switching into LRS can excessively decrease the resistance in LRS [4, 7]. In addition, the electrical charges accumulated in the capacitance component in the device can lower the resistance in LRS [8]. In several reports on RRAMs with the reactive element electrodes such as Ti and Ta, the lowering of the current during the reset process has been demonstrated [4, 9] and the relation between the reset current reduction and the oxygen gettering ability of electrode materials has been discussed [9, 10]. However, the further accurate control of the resistance in LRS is thought to be performed with the embedded transistor in series with the RRAM cell [11-13]. Since the present device consists of only the RRAM cell without the transistor, the measured operation current is regarded as being extrinsic and the detailed comparison of such current magnitude is beyond the scope of the present paper.

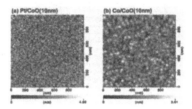

Figure 1. AFM images of (a) Pt/CoO(10nm) and (b) Co/CoO(10nm). The scanned area in those image is 1×1 μm^2.

In order to regulate the excessive current during the switching into LRS (i.e. forming and set processes), the current compliance program of the semiconductor parameter analyzer was used for PP, PW, CP, and CW, while the load resistor in series with RRAM was used for PT.

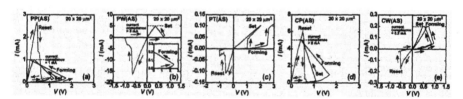

Figure 2. Resistance switching characteristics observed in (a) PP, (b) PW, (c) PT, (d) CP, (e) CW in their as prepared states.

We have evaluated the robustness to PDA by comparing the initial resistance in the as prepared state and that after PDA. In this experiment, the device in the identical address was used in order to understand precisely the influence of PDA. For Figs. 3(a)-3(e), the transverse axis is the initial resistance in the as prepared state (R_{ini}^{AS}) of PP, PW, PT, CP, and CW, respectively, while the longitudinal axis is the initial resistance after PDA (R_{ini}^{PDA}) of those devices. The values of the read out voltage V_{Read} and the device size were also shown in those figures. As observed in Figs. 3(a), 3(c), and 3(d), the CoO RRAM device with Pt and Ti showed the severe decrease of the initial resistance since $R_{ini}^{PDA} \ll R_{ini}^{AS}$. On the other hands, R_{ini}^{PDA} was comparable with R_{ini}^{AS} in PW and CW.

Resistance switching characteristics can still be confirmed in PP, PW, and CW, whereas it disappeared in PT and CP. The typical I-V curves in PP, PW, PT, CP, and CW were shown in Figs. 4(a) – 4(e). According to Fig. 3(a), the resistance of PP was significantly decreased by PDA. The resistance switching in PP after PDA can be categorized into two groups. In group-A, no resistance switching was observed. On the other hands, in group-B, the reset process was observed in the first voltage sweep after PDA [Fig. 4(a)]. PW and CW after PDA switched from a highly resistive initial state to LRS and they consecutively switched into HRS [Figs. 4(b) and 4(e), respectively]. Finally, the switching from HRS to LRS was also observed in PW and CW. Therefore, the forming reset, and set processes were clearly observed in PW and CW after PDA. In PT and CP, no noticeable current hysterias corresponding to the resistance switching was observed [Figs. 4(c) and 4(d)].

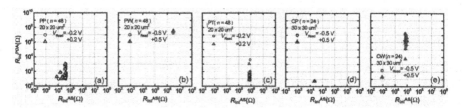

Figure 3. The relation between the initial resistance in the as prepared state (R_{ini}^{AS}) and that in the annealed state (R_{ini}^{PDA}) for (a) PP, (b) PW, (c) PT, (d) CP, and (e) CW. V_{Read} and n denote the read out voltage and the number of the examined devices.

The influence of PDA on the initial resistance and resistance switching characteristics is summarized in Table 1. The significant decrease of the initial resistance value in PT can be understood as the oxygen gettering effect of Ti with respect to the affinity with oxygen. The reduction by Ti in several oxides has been clearly observed even in the oxides of Si, Zr, and Hf [14, 15] which have much lower oxygen potential compared with Co. In addition to that, since the lowering of the initial resistance was observed in PP, it can be reasonably expected that the PDA process in the Ar atmosphere basically reduces the CoO layer. It is expected that this reduction inhomogeneously decreases the resistance in CoO because the reset process was observed in a part of PP after PDA. Since there is a concentration gradient across the Co/CoO interface in CP can further promote the reduction compared to PP. Consequently, in CP, being different from PP, the reset process in the first voltage sweep was no longer observed. For W, it has been reported that the solid solution oxygen stably exist, which is observed in several metallic elements [16]. The influence of oxygen in W on the device properties has been discussed in the high-k gate stack structure of HfO$_2$/W [17]. In PW and CW, the values of the initial resistance are slightly increased by PDA. In consideration for the oxygen partial pressure dependence of the CoO resistivity, a moderate reduction in CoO is indicated. Probably because of the solid solution oxygen in W, the reduction process is partially suppressed in PW and CW, leading to the relatively lager value of the initial resistance compared with PP, PT, and CP.

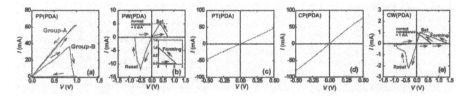

Figure 4. *I-V* curves in (a) PP, (b) PW, (c) PT, (d) CP, and (e) CW measured after PDA. The inset in Fig. 4(b) is the forming process in PW. The current compliance program was used in the forming and set processes in PW and CW. The arrows in figures correspond to the direction of the voltage sweep during measurements.

Table 1. Summary of influence of PDA on the initial resistance (R_{ini}) and resistance switching (RS) characteristics in CoO RRAM with various electrode materials.

	PP	PW	PT	CP	CW
BE	Pt	Pt	Pt	Co	Co
TE	Pt	W	Ti	Pt	W
R_{ini} after PDA	Low	High	Low	Low	High
RS after PDA	Reset in the 1st voltage sweep is observed	Forming, reset, and set are observed	No RS	No RS	Forming, reset, and set are observed
Dominant origin of R_{ini} change	Reduction in the reducing Ar ambient	Oxygen supplied from W	Reduction by Ti	Reduction by Co	Oxygen supplied from W

CONCLUSIONS

The electrode material dependence of the CoO layer thermal stability during the high temperature annealing process in the reducing ambient was systematically investigated. The reducing Ar ambient promotes the quite local reduction of CoO because the reset process was observed in the first voltage sweep in the Pt/CoO/Pt device after the post deposition annealing (PDA) process. In addition to the influence of the annealing ambient, the gettering and supplying processes of oxygen to/from electrode materials play a critical role on the thermal stability of CoO. When the Ti electrode is used, the reduction due to the oxygen gettering effect of Ti is indicated because the initial resistance in Pt/CoO/Ti device is severely decreased. The reduction of the CoO layer is also expected when the Co electrode is used because the diffusion of oxygen into Co can be promoted by the concentration gradient at the Co/CoO interface in the Co/CoO/Pt device. On the other hands, the W electrodes which is thought to have the solid-solution oxygen can stabilize the CoO layer during PDA although W is more affinitive oxygen compared with Co. It can be suggested that the oxygen delivery at the electrode /oxide layer interface is a critical factor in designing the thermally stable stacking structure for resistance random access memory.

REFERENCES

1 S. Seo, M. J. Lee, D. H. Seo, E. J. Jeoung, D.-S. Suh, Y. S. Joung, I. K. Yoo, I. R. Hwang, S. H. Kim, I. S. Byun, J.-S. Kim. J. S. Choi, and B. H. Park, Appl. Phys. Lett. **85**, 5655 (2004).

2 K. Kinoshita, T. Tamura, M. Aoki, Y. Sugiyama, and H. Tanaka, Appl. Phys. Lett. **89**, 103509 (2006).

3 B. J. Choi, D. S. Jeong, S. K. Kim, C. Rohde, S. Choi, J. H. Oh, H. J. Kim, C. S. Hwang, K. Szot, R. Waser, B. Reichenberg, and S. Tiedke, J. Appl. Phys. **98**, 033715 (2005).

4 H. Shima, F. Takano, H. Muramatsu, H. Akinaga, Y. Tamai, I. H. Inoue, and H. Takagi, Appl. Phys. Lett. **93**, 113504 (2008).

5 Y. H. Do, J. S. Kwak, J. P. Hong, K. Jung, and H. Y. Im, J. Appl. Phys. **104**, 114512 (2008).

6 C.-Y. Lin, C.-Y. Wu, C.-Y. Wu, T.-Y. Tseng, and C. Hu, J. Appl. Phys. **102**, 094101 (2007).

7 L. Goux, J. G. Lisoni, X. P. Wang, M. Jurczak, and D. J. Wouters, IEEE Trans. Electron Devices **56**, 2363 (2009).

8 K. Kinoshita, T. Tsunoda, Y. Sato, H. Noshiro, S. Yagaki, M. Aoki, and Y. Sugiyama, Appl. Phys. Lett. **93**, 033506 (2008).

9 H. Y. Lee, Y. S. Chen, P. S. Chen, T. Y. Wu, F. Chen, C. C. Wang, P. J. Tzeng, M.-J. Tsai, and C. Lien, IEEE Electron Device Lett. **31**, 44 (2010).

10 H. Y. Lee, P.-S. Chen, T.-Y. Wu, Y. S. Chen, F. Chen, C.-C. Wang, P.-J. Tzeng, C. H. Lin, M.-J. Tsai, and C. Lien, IEEE Electron Device Lett. **30**, 703 (2009).

11 A. Chen, S. Haddad, Y.-C. Wu, T.-N. Fang, Z. Lan, S. Avanzine, S. Pangrle, M. Buynoski, M. Rathor, W. Chai, N. Tripsas, C. Bill, M. VanBuskirk, and M. Taguchi, IEDM Tech. Dig., 2005, pp. 765-768.

12 K. Tsunoda, K. Kinoshita, H. Noshiro, Y. Yamazaki, T. Iizuka, Y. Ito, A. Takahashi, A. Okano, Y. Sato, T. Fukano, M. Aoki, and Y. Sugiyama, IEDM Tech. Dig., 2007, pp. 637-640

13 H. Y. Lee, P. S. Chen, T. Y. Wu, Y. S. Chen, C. C. Wang, P. J. Tzeng, C. H. Lin, F. Chen, C. H. Lien, and M.-J. Tsai, IEDM Tech. Dig., 2008, pp. 297-300.

14 K. Nakajima, A. Fujiyoshi, Z. Ming, M. Suzuki, and K. Kimura, J. Appl. Phys. **102**, 064507 (2007).

15 H. Kim, P. C. McIntyre, C. O. Chui, and K. C. Saraswat, and S. Stemmer, J. Appl. Phys. **96**, 3467 (2004).

16 V. P. Kobyakov, and V. I. Ponomarev, Crystallography Reports **47**, 114 (2002).

17 E. J. Preisler, S. Guha, M. Copel, N. A. Bojarczuk, M. C. Reuter, and E. Gusev, Appl. Phys. Lett. **85**, 6230 (2004).

AUTHOR INDEX

SUBJECT INDEX

Printed in the United States
by Baker & Taylor Publisher Services